Uncertainty-Based Information

Studies in Fuzziness and Soft Computing

Editor-in-chief
Prof. Janusz Kacprzyk
Systems Research Institute
Polish Academy of Sciences
ul. Newelska 6
01-447 Warsaw, Poland
E-mail: kacprzyk@ibspan.waw.pl

Vol. 3. A. Geyer-Schulz
*Fuzzy Rule-Based Expert Systems and
Genetic Machine Learning, 2nd ed. 1996*
ISBN 3-7908-0964-0

Vol. 4. T. Onisawa and J. Kacprzyk (Eds.)
*Reliability and Safety Analyses under
Fuzziness, 1995*
ISBN 3-7908-0837-7

Vol. 5. P. Bosc and J. Kacprzyk (Eds.)
*Fuzziness in Database Management
Systems, 1995*
ISBN 3-7908-0858-X

Vol. 6. E.S. Lee and Q. Zhu
Fuzzy and Evidence Reasoning, 1995
ISBN 3-7908-0880-6

Vol. 7. B.A. Juliano and W. Bandler
Tracing Chains-of-Thought, 1996
ISBN 3-7908-0922-5

Vol. 8. F. Herrera and J.L. Verdegay (Eds.)
*Genetic Algorithms and Soft Computing,
1996*, ISBN 3-7908-0956-X

Vol. 9. M. Sato et al.
*Fuzzy Clustering Models and Applications,
1997*, ISBN 3-7908-1026-6

Vol. 10. L.C. Jain (Ed.)
*Soft Computing Techniques in Knowledge-
based Intelligent Engineering Systems,
1997,*
ISBN 3-7908-1035-5

Vol. 11. W. Mielczarski (Ed.)
*Fuzzy Logic Techniques in Power Systems,
1998*, ISBN 3-7908-1044-4

Vol. 12. B. Bouchon-Meunier (Ed.)
*Aggregation and Fusion of Imperfect
Information, 1998*
ISBN 3-7908-1048-7

Vol. 13. E. Orłowska (Ed.)
*Incomplete Information:
Rough Set Analysis, 1998*
ISBN 3-7908-1049-5

Vol. 14. E. Hisdal
*Logical Structures for Representation
of Knowledge and Uncertainty, 1998*
ISBN 3-7908-1056-8

Vol. 15. G.J. Klir and M.J. Wierman
*Uncertainty-Based Information,
2nd ed. 1999*
ISBN 3-7908-1242-0

Vol. 16. D. Driankov and R. Palm (Eds.)
Advances in Fuzzy Control, 1998
ISBN 3-7908-1090-8

Vol. 17. L. Reznik, V. Dimitrov and
J. Kacprzyk (Eds.)
Fuzzy Systems Design, 1998
ISBN 3-7908-1118-1

Vol. 18. L. Polkowski and A. Skowron
(Eds.)
*Rough Sets in Knowledge Discovery 1,
1998*, ISBN 3-7908-1119-X

Vol. 19. L. Polkowski and A. Skowron
(Eds.)
*Rough Sets in Knowledge Discovery 2,
1998*, ISBN 3-7908-1120-3

Vol. 20. J.N. Mordeson and P.S. Nair
Fuzzy Mathematics, 1998
ISBN 3-7908-1121-1

Vol. 21. L.C. Jain and T. Fukuda (Eds.)
*Soft Computing for Intelligent Robotic
Systems, 1998*
ISBN 3-7908-1147-5

Vol. 22. J. Cardoso and H. Camargo (Eds.)
Fuzziness in Petri Nets, 1999
ISBN 3-7908-1158-0

Vol. 23. P.S. Szczepaniak (Ed.)
*Computational Intelligence and
Applications, 1999*
ISBN 3-7908-1161-0

Vol. 24. E. Orłowska (Ed.)
Logic at Work, 1999
ISBN 3-7908-1164-5

continued on page 170

George J. Klir · Mark J. Wierman

Uncertainty-Based Information

Elements of Generalized Information Theory

Second Corrected Edition

With 11 Figures
and 10 Tables

Physica-Verlag

A Springer-Verlag Company

Professor George J. Klir
Center for Intelligent Systems
and
Department of Systems Science and Industrial Engineering
Thomas J. Watson School of Engineering and Applied Science
Binghamton University – SUNY
Binghamton, NY 13902-600
USA

Professor Mark J. Wierman
Center for Research in Fuzzy Mathematics and Computer Science
and
Mathematics and Computer Science Department
Creighton University
Omaha, NE 68178-2090
USA

ISBN 3-7908-1242-0 Physica-Verlag Heidelberg New York
ISBN 3-7908-1073-8 1st Ed. Physica-Verlag Heidelberg New York

Cataloging-in-Publication Data applied for
Die Deutsche Bibliothek – CIP-Einheitsaufnahme
Uncertainty-based information: elements of generalized information theory: with 11 figures
and 10 tables / George J. Klir; Mark J. Wierman. – Heidelberg; New York: Physica-Verl.,
1999
   (Studies in fuzziness and soft computing; Vol. 15 2nd edition)
   ISBN 3-7908-1242-0

Hardcover Design: Erich Kirchner, Heidelberg

SPIN 10741828   88/2202-5 4 3 2 1 0 – Printed on acid-free paper

PREFACE

Information is precious. It reduces our uncertainty in making decisions. Knowledge about the outcome of an uncertain event gives the possessor an advantage. It changes the course of lives, nations, and history itself.

Information is the food of Maxwell's demon. His power comes from knowing which particles are hot and which particles are cold. His existence was paradoxical to classical physics and only the realization that information too was a source of power led to his taming.

Information has recently become a commodity, traded and sold like orange juice or hog bellies. Colleges give degrees in information science and information management. Technology of the computer age has provided access to information in overwhelming quantity. Information has become something worth studying in its own right.

The purpose of this volume is to introduce key developments and results in the area of generalized information theory, a theory that deals with uncertainty-based information within mathematical frameworks that are broader than classical set theory and probability theory. The volume is organized as follows.

First, basic concepts and properties of relevant theories of uncertainty are introduced in Chapter 2. This background knowledge is employed in Chapter 3, whose aim is to give a comprehensive overview of conceptual and mathematical issues regarding measures of the various types of uncertainty in the introduced theories. The utility of these uncertainty measures is then exemplified by three broad principles of uncertainty in Chapter 4. In Chapter 5, current results regarding generalized information theory are critically assessed, some important open problems are overviewed, and prospective future directions on research in this area are discussed. Finally, a complicated proof of one theorem stated in the text is presented in the Appendix.

ACKNOWLEDGEMENTS

Most of the original results presented in this monograph were obtained over the period of about 15 years (1981-1996) at Binghamton University (SUNY-Binghamton) by, or under the supervision of, George J. Klir. The results are based on the work of Roger E. Cavallo, David Harmanec, Masahiko Higashi, George J. Klir, Matthew Mariano, Yin Pan, Behzad Parviz, Arthur Ramer, Ute St. Clair, Mark J. Wierman, Zhenyuan Wang, and Bo Yuan (see appropriate references in the bibliography). This work has been supported, in part, by grants from the National Science Foundation (No. IST85-44191, 1985-88, and No. IRI90-15675, 1991-93), the Office of Naval Research (No. N00014-94-1-0263, 1994-96), and the Rome Laboratory of the Air Force (No. F30602-94-1-0011, 1994-97). The support of these research-sponsoring agencies is gratefully acknowledged.

We are also grateful for the invitation from the Center for Research in Fuzzy Mathematics and Computer Science at Creighton University in Omaha, Nebraska, to publish an earlier version of this monograph in the Lecture Notes in Fuzzy Mathematics and Computer Science series. Without the support of John N. Mordeson, Director of the Research Center, and Michael Proterra, S.J., Dean of the College of Arts and Sciences, both at Creighton University, this book would not have been possible.

CONTENTS

LIST OF FIGURES

LIST OF TABLES

LIST OF SYMBOLS

$\lvert A \rvert$	scalar cardinality (sigma count) of A
$A, B, \ldots$	fuzzy set membership functions
$\bar{A}$	standard fuzzy complement of A
A^{co}	general fuzzy complement of A
$^{\alpha}A$	α-cut of A
$^{\alpha+}A$	strong α-cut of A
AU	aggregate evidential uncertainty
$\mathcal{B}$	set of belief measures
Bel	belief measure
C	confusion
$card(A)$	fuzzy cardinality of A
D	discord
E	dissonance
$\emptyset$	empty set
$\mathcal{F}$	focal set
f	fuzziness
^{e}f	fuzzy entropy
g	a fuzzy measure
H	Hartley measure of uncertainty
HL	Hartley-like measure of uncertainty
$h(A)$	height of A
$\Lambda(A)$	level set of A
L	lattice
m	basic probability assignment
$\mathbb{N}$	the set of natural numbers

$\mathbb{N}_n$	the set $\{1, 2, \ldots, n\}$
N	nonspecificity
Nec	necessity measure
NS	nonspecificity plus strife
Pl	plausibility measure
Pos	possibility measure
$\tilde{P}(X)$	fuzzy power set of X
$\mathcal{P}(X)$	power set of X
Pro	probability measure
P_*	lower probability
P^*	upper probability
$\mathcal{P}$	set of probability measures
p	probability distribution
$\mathcal{R}$	set of possibility measures
$\mathbb{R}$	the set of real numbers
$\mathbb{R}^+$	the set of nonnegative real numbers
r	possibility distribution function
$\mathbf{r}$	possibility distribution
S	Shannon entropy
ST	strife
$sub(A, B)$	the degree of subsethood of A in B
U	U-uncertainty
X	universal set
$\langle x, y \rangle$	ordered pair
$X \uparrow Z$	cylindric extension
$\hat{Z}$	cylindric closure

1
INTRODUCTION

For three hundred years (from about the mid-seventeenth century, when the formal concept of numerical probability emerged, until the 1960s), uncertainty was conceived solely in terms of probability theory. This seemingly unique connection between uncertainty and probability is now challenged. The challenge comes from several mathematical theories, distinct from probability theory, which are demonstrably capable of characterizing situations under uncertainty. The most visible of these theories, which began to emerge in the 1960s, are the theory of fuzzy sets [Zadeh, 1965], evidence theory [Dempster, 1967a,b; Shafer, 1976], possibility theory [Zadeh, 1978], and the theory of fuzzy measures [Sugeno, 1974, 1977].

When the question of measuring uncertainty within these theories was investigated (mostly in the 1980s), it became clear that there are several distinct types of uncertainty. That is, it was realized that uncertainty is a multidimensional concept. Which of its dimensions are actually manifested is determined by the mathematical theory employed. The multidimensional nature of uncertainty was obscured when uncertainty was conceived solely in terms of probability theory, in which it is manifested by only one of its dimensions.

Well justified measures of uncertainty of relevant types are now available not only in the classical set theory and probability theory, but also in the theory of fuzzy sets, possibility theory, and evidence theory. The purpose of this book is to overview these measures and explain how they can be utilized for measuring information. In addition, three methodological principles based upon these measures are described: a principle of minimum uncertainty, a principle of maximum uncertainty, and a principle of uncertainty invariance.

1.1 Significance of Uncertainty

When dealing with real-world problems, we can rarely avoid uncertainty. At the empirical level, uncertainty is an inseparable companion of almost any measurement, resulting from a combination of inevitable measurement errors and resolution limits of measuring instruments. At the cognitive level, it emerges from the vagueness and ambiguity inherent in natural language. At the social level, uncertainty has even strategic uses and it is often created and maintained by people for different purposes (privacy, secrecy, propriety). As a result of the famous Gödel theorem, we are now aware that, surprisingly, even mathematics is not immune from uncertainty. A comprehensive and thorough coverage of these and many other facets of uncertainty in human affairs was prepared by Smithson [1989].

When encountered with uncertainty regarding our purposeful actions, we are forced to make decisions. This subtle connection between uncertainty and decision making is explained with great clarity by Shackle [1961]:

> In a predestinate world, decision would be illusory; in a world of a perfect foreknowledge, *empty*; in a world without natural order, *powerless*. Our intuitive attitude to life implies non-illusory, non-empty, non-powerless decision. ... Since decision in this sense excludes both perfect foresight and anarchy in nature, it must be defined as choice in face of bounded uncertainty.

This quote captures well the significance of uncertainty in human affairs. Indeed, conscious decision making, in all its varieties, is perhaps the most fundamental capability of human beings. In order to understand this capability and being able to improve it, we must understand the notion of uncertainty first. This requires, in turn, that the various facets of uncertainty be adequately conceptualized and formalized in appropriate mathematical languages.

As argued by Smithson [1989], "Western intellectual culture has been preoccupied with the pursuit of absolutely certain knowledge or, barring that, the nearest possible approximation of it." This preoccupation appears to be responsible not only for the neglect of uncertainty by Western science, but also, prior to the 1960s, the absence of adequate conceptual frameworks for seriously studying it.

What caused the change of attitude towards uncertainty since the 1960s? It seems that this change was caused by a sequence of events. The emergence of computer technology in the 1950s opened new methodological possibilities. These, in turn, aroused interest of some researchers to study certain problems that, due to their enormous complexities, were previously beyond the scope of scientific inquiry. In his insightful paper, Warren Weaver [1948] refers to them as problems of *organized complexity*. These problems involve nonlinear systems with large numbers of components and rich interactions among the components, which are usually nondeterministic, but not as a result of randomness that could yield meaningful statistical averages. They are typical in life, behavioral, social, and environmental sciences, as well as in applied fields such as modern technology or medicine.

The major difficulty in dealing with problems of organized complexity is that they require massive combinatorial searches. This is a weakness of the human mind while, at the same time, it is a strength of the computer. This explains why advances into the realm of organized complexity have been closely correlated with advances in computer technology.

Shortly after the emergence of computer technology, it was the common belief of many scientists that the level of complexity we can handle is basically a matter of the level of the computational power at our disposal. Later, in the early 1960s, this naive belief was replaced with a more realistic outlook. We began to understand that there are definite limits in dealing with complexity, which neither our human capabilities nor any computer technology can overcome. One such limit was determined by Hans Bremermann [1962] by simple considerations based on quantum theory. The limit is expressed by the proposition: "No data processing systems, whether artificial or living, can process more than 2×10^{47} bits per second per gram of its mass." To process a certain number of bits means, in this statement, to transmit that many bits over one or several channels within the computing system.

Using the limit of information processing obtained for one gram of mass and one second of processing time, Bremermann then calculates the total number of bits processed by a hypothetical computer the size of the Earth within a time period equal to the estimated age of the Earth. Since the mass and age of the Earth are estimated to be less than 6×10^{27} grams and 10^{10} years, respectively, and each year contains approximately 3.14×10^7 seconds, this imaginary computer would not be able to process more than 2.56×10^{92} bits or, when rounding up to the nearest power of ten, 10^{93} bits. The last number — 10^{93} — is usually referred to as Bremermann's limit and problems that require processing more than 10^{93} bits of information are called *transcomputational problems*.

Bremermann's limit seems at first sight rather discouraging, even though it is based on overly optimistic assumptions (more reasonable assumptions would result in a number smaller than 10^{93}). Indeed many problems dealing with systems of even modest size exceed the limit in their information processing demands. The nature of these problems have been extensively studied within an area referred to as the *theory of computational complexity*

[Garey and Johnson, 1979], which emerged in the 1960s as a branch of the general theory of algorithms.

In spite of the insurmountable computational limits, we continue to pursue the many problems that possess the characteristics of organized complexity. These problems are too important for our well being to give up on them. The main challenge in pursuing these problems narrows down fundamentally to one question: how to deal with systems and associated problems whose complexities are beyond our information processing limits? That is, how can we deal with these systems and problems if no computational power alone is sufficient? The only possible answer is that we must adequately simplify them to make them computationally tractable. In every simplification, unfortunately, we are doomed to loose something.

In general, we deal with problems in terms of systems that are constructed either as models of some aspects of reality or as models of some desirable man-made objects. The purpose of constructing models of the former type is to understand some phenomena of reality, be it natural or man-made, making adequate predictions or retrodictions, learning how to control the phenomena in any desirable way, and utilizing all these capabilities for various ends; models of the latter type are constructed for the purpose of prescribing operations by which a conceived artificial object can be constructed in such a way that desirable objective criteria are satisfied within given constraints.

In constructing a model, we always attempt to maximize its usefulness. This aim is closely connected with the relationship among three key characteristics of every systems model: *complexity, credibility,* and *uncertainty*. This relationship, which is a subject of current study in systems science [Klir, 1985, 1991a], is not as yet fully understood. We only know that uncertainty has a pivotal role in any efforts to maximize the usefulness of systems models. Although usually (but not always) undesirable when considered alone, uncertainty becomes very valuable when considered in connection to the other characteristics of systems models: a slight increase in uncertainty may often significantly reduce complexity and; at the same time, increase credibility of the model. Uncertainty is thus an important commodity in the modelling business, a commodity which can be traded for gains in the other essential characteristics of models. It is this important positive role of uncertainty, we believe, which is primarily responsible for the rapidly growing interest, during the last three decades or so, in investigating uncertainty in all its manifestations.

As argued by Klir [1995a], the currently ongoing change in attitudes towards uncertainty has all of the features characterizing a paradigm shift in science, a notion introduced by Kuhn [1962].

1.2 Uncertainty and Information

The concept of information, as a subject of this book, is intimately connected with the concept of uncertainty. The most fundamental aspect of this connection is that the uncertainty involved in any problem-solving situation is a result of some information deficiency. Information (pertaining to the model within which the situation is conceptualized) may be incomplete, imprecise, fragmentary, not fully reliable, vague, contradictory, or deficient in some other way. In general, these various information deficiencies may result in different types of uncertainty.

Assume that we can measure the amount of uncertainty involved in a problem solving situation conceptualized in a particular mathematical theory. Assume further that the amount of uncertainty can be reduced by obtaining relevant information as a result of some action (finding a relevant new fact, designing a relevant experiment and observing the experimental outcome, receiving a requested message, discovering a relevant historical record). Then, the amount of information obtained by the action may be measured by the reduction of uncertainty that results from the action.

Information measured solely by the reduction of uncertainty does not capture the rich notion of information in human communication and cognition. It is not explicitly concerned with semantic and pragmatic aspects of information viewed in the broader sense. This does not mean, however, that information viewed in terms of uncertainty reduction ignores these aspects. It does not ignore them, but they are assumed to be fixed in each particular application. Furthermore, as argued by Dretske [1981, 1983], the notion of information as uncertainty reduction is sufficiently rich as a base for additional treatment, treatment through which human communication and cognition can adequately be explicated.

It should be noted at this point that the concept of information has also been investigated in terms of the theory of computability, independent of the concept of uncertainty. In this approach, the amount of information represented by an object is measured by the length of the shortest possible program written in some standard language (e.g., a program for the standard Turing machine) by which the object is described in the sense that it can be computed. Information of this type is usually referred to as *descriptive information* or *algorithmic information* [Kolmogorov, 1965; Chaitin, 1987].

Some additional approaches to information have recently appeared in the literature. For example, Devlin [1991] formulates and investigates information in terms of logic, while Stonier [1990] views information as a physical property defined as the capacity to organize a system or to maintain it in an organized state.

As already mentioned, this book is concerned solely with information conceived in terms of uncertainty reduction. To distinguish this conception of information from various other conceptions of information, let us call it *uncertainty–based information*.

The nature of uncertainty-based information depends on the mathematical theory within which uncertainty pertaining to various problem-solving situations is formalized. Each formalization of uncertainty in a problem-solving situation is a mathematical model of the situation. When we commit ourselves to a particular mathematical theory, our modelling becomes necessarily limited by the constraints of the theory. Clearly, a more general theory is capable of capturing uncertainties of some problem situations more faithfully than its less general competitors. As a rule, however, it involves greater computational demands.

Uncertainty-based information was first conceived in terms of classical set theory by Hartley [1928] and, later, in terms of probability theory by Shannon [1948]. The term *information theory* has almost invariably been used to refer to a theory based upon a measure of probabilistic uncertainty established by Shannon [1948]. The Harley measure of information [Hartley, 1928] has usually been viewed as a special case of the Shannon measure. This view, which is flawed (as explained in Sec. 5.1), was responsible for a considerable conceptual confusion in the literature.

Research on a broader conception of uncertainty-based information, liberated from the confines of classical set theory and probability theory, began in the early 1980s [Higashi and Klir, 1982, 1983; Höhle, 1982; Yager, 1983]. The name *generalized information theory* was coined for a theory based upon this broader conception [Klir, 1991b].

The ultimate goal of generalized information theory is to capture the properties of uncertainty-based information formalized within any feasible mathematical framework. Although this goal has not been fully achieved as yet, substantial progress has been made in this direction since the early 1980s. In addition to classical set theory and probability theory, uncertainty-based information is now well understood in fuzzy set theory, possibility theory, and evidence theory.

2
UNCERTAINTY FORMALIZATIONS

Mathematical theories of uncertainty in which measures of uncertainty are now well established are fuzzy set theory, evidence theory, and possibility theory, in addition to classical set theory and probability theory. Some of these theories are connected. Classical set theory is subsumed under fuzzy set theory. Probability theory and possibility theory are branches of evidence theory, while evidence theory is, in turn, a branch of fuzzy measure theory. Although information aspects of fuzzy measure theory have not been investigated as yet, the theory is briefly introduced in this chapter because it represents a broad framework for future research.

Fuzzy set theory can be combined with fuzzy measure theory and its various branches. This combination is referred to as *fuzzification*. All the established measures of uncertainty can be readily extended to their fuzzified counterparts.

The coverage of each mathematical theory of uncertainty in this chapter is not fully comprehensive. Covered are only those aspects of each theory that are essential for understanding issues associated with measuring uncertainty in that theory. However, the reader is provided with references to publications that cover the theories in comprehensive and up-to-date fashion.

2.1   Classical Sets: Terminology and Notation

A *set* is a collection of objects called elements. Conventionally, capital letters represent sets and small letters represent elements. Symbolically, the statement "5 is an *element* of set A" is written as $5 \in A$.

A set is defined using one of three methods. In the first method the elements of the set are explicitly listed, as in

$$A = \{1, 3, 5, 7, 9\}. \tag{2.1}$$

The second method for defining a set is to give a rule or property that a potential element must posses in order to be included in the set. An example of this is the set

$$A = \{x \mid x \text{ is an odd number between zero and ten}\}, \tag{2.2}$$

where | means "such that." This is the same set A that was defined explicitly by listing its elements in Eq. (2.1). Both of these methods of defining a set assume the existence of a universal set that contains all the objects of discourse; this universal set is usually denoted in this book by X. Some common universal sets in mathematics have standardized symbols to represent them, such as, $\mathbb{N}$ for the natural numbers, $\mathbb{Z}$ for the integers, and $\mathbb{R}$ for the real numbers. The third way to specify a set is through a *characteristic function*. If χ_A is the characteristic function of a set A, then χ_A is a function from the universe of discourse X to the set $\{0, 1\}$, where

$$\chi_A(x) = \begin{cases} 1 & \text{if } x \text{ is an element of } A \\ 0 & \text{if } x \text{ is not an element of } A. \end{cases} \tag{2.3}$$

For the set A of odd natural numbers less than ten the characteristic function is

$$\chi_A(x) = \begin{cases} 1 & x = 1, 3, 5, 7, 9 \\ 0 & \text{otherwise}. \end{cases} \tag{2.4}$$

A set A is contained in or equal to another set B, written $A \subseteq B$, if every element of A is an element of B, that is, if $x \in A$ implies that $x \in B$. If A is contained in B, then A is said to be a *subset* of B, and B is said to be a *superset* of A. Two sets are equal, symbolically $A = B$, if they contain exactly the same elements; therefore, if $A \subseteq B$ and $B \subseteq A$ then $A = B$. If $A \subseteq B$ and A is not equal to B, then A is called a proper subset of B, written $A \subset B$. The negation of each of these propositions is expressed symbolically by a slash crossing the operator. That is $x \notin A$, $A \nsubseteq B$ and $A \neq B$ represent, respectively, x is not an element of A, A is not a subset of B, and A is not equal to B.

The *intersection* of two sets is a new set that contains every object that is simultaneously an element of both the set A and the set B. If $A = \{1, 3, 5, 7, 9\}$ and $B = \{1, 2, 3, 4, 5\}$, then the intersection of set A and

B is the set $A \cap B = \{1, 3, 5\}$. The *union* of the two sets contains all the elements of either set A or set B. With the sets A and B defined previously $A \cup B = \{1, 2, 3, 4, 5, 7, 9\}$.

The *complement* of a set A, denoted $\bar{A}$, is the set of all elements of the universal set that are not elements of A. With $A = \{1, 3, 5, 7, 9\}$ and the universal set $X = \{1, 2, 3, 4, 5, 6, 7, 8, 9\}$, the complement of A is $\bar{A} = \{2, 4, 6, 8\}$. One last set operation is *set difference*, $A - B$, which is the set of all elements of A that are not elements of B. With A and B as defined previously, $A - B = \{7, 9\}$. The complement of A is equivalent to $X - A$.

All of the concepts of set theory can be recast in terms of the *characteristic functions* of the sets involved. For example we have that $A \subseteq B$ if and only if $\chi_A(X) \leq \chi_B(X)$ for all $x \in X$. The phrase "for all" occurs so often in set theory that a special symbol is used as an abbreviation; $\forall$ represents the phrase "for all". Similarly the phrase "there exists" is abbreviated $\exists$. For example, the definition of set equality can be restated as: $A = B$ if and only if $\forall x \in X$, $\chi_A(x) = \chi_B(x)$.

Sets can be finite or infinite. A finite set is one that contains a finite number of elements. The size of a finite set, called its cardinality, is the number of elements it contains. If $A = \{1, 3, 5, 7, 9\}$, then the cardinality of A, or $|A|$, is 5. A set may be empty, that is, it may contain no elements. The *empty set* is given a special symbol $\emptyset$; thus $\emptyset = \{\}$ and $|\emptyset| = 0$.

The set of all subsets of a given set X is called the *power set* of X and is denoted $\mathcal{P}(X)$. If X is finite and $|X| = n$ then the number of subsets of X is 2^n.

The ordered pair formed by two objects x and y, where $x \in X$ and $y \in Y$, is denoted by $\langle x, y \rangle$. The set of all ordered pairs, where the first element is contained in a set X and the second element is contained in a set Y, is called the Cartesian product of X and Y and is denoted as $X \times Y$. If $X = \{1, 2\}$ and $Y = \{a, b\}$ then $X \times Y = \{\langle 1, a \rangle, \langle 1, b \rangle, \langle 2, a \rangle, \langle 2, b \rangle\}$. Note that the size of $X \times Y$ is the product of the size of X and the size of Y when X and Y are finite: $|X \times Y| = |X| \cdot |Y|$.

If $R \subseteq X \times Y$, then we call R a relation between X and Y. If $\langle x, y \rangle \in R$, then we also write $x \, R \, y$ to signify that x is related to y by R. A function is a mapping from a set X to a set Y, denoted $f : X \rightarrow Y$. We write $f(x) = y$ to indicate that the function f maps $x \in X$ to $y \in Y$.

A type of sets that is used extensively in this text is a bounded subset of the natural numbers. The set $\mathbb{N}_n$ will designate the natural numbers less than or equal to some given natural number n. Thus

$$\mathbb{N}_n = \{x \in \mathbb{N} \mid x \leq n\}. \tag{2.5}$$

For example $\mathbb{N}_6 = \{1, 2, 3, 4, 5, 6\}$.

2.2 Fuzzy Set Theory

Fuzzy set theory is an outgrowth of classical set theory. Contrary to the classical concept of a set, or *crisp set,* the boundary of a *fuzzy set* is not precise. That is, the change from nonmembership to membership in a fuzzy set is gradual rather than abrupt. This gradual change is expressed by a *membership function.* Two distinct notations are most commonly employed in the literature to denote membership functions of fuzzy sets. In one of these, the membership function of a fuzzy set A is denoted by μ_A and its form is usually

$$\mu_A : X \to [0,1], \tag{2.6}$$

where X denotes the crisp universal set under consideration and A is a label of the fuzzy set defined by this function. The value $\mu_A(x)$ expresses for each $x \in X$ the *grade of membership* of element x of X in fuzzy set A or, in other words, the *degree of compatibility* of x with the concept represented by the fuzzy set A. These values are in the interval between zero and one inclusive.

In the second notation, the distinction between the symbol A, denoting the fuzzy set, and the symbol μ_A, denoting the membership function of A, is not made. That is, the membership function of a fuzzy set A is denoted by the same symbol A, where

$$A : X \to [0,1], \tag{2.7}$$

and $A(x)$ is the degree of membership of x in A for each $x \in X$. No ambiguity results from this double use of the same symbol since each fuzzy set is uniquely defined by one particular membership function.

In this book, the second notation is adopted. It is simpler and, by and large, more popular in current literature on fuzzy set theory. Since classical sets are viewed from the standpoint of fuzzy set theory as special fuzzy sets, often referred to as *crisp sets*, the same notation is used for them. Moreover, we use the symbol $\tilde{\mathcal{P}}(X)$ to denote the set of all fuzzy subsets of X (the fuzzy power set of X).

An example of membership grade functions defining five fuzzy sets on the closed interval $X = [0, 100]$ of real numbers is shown in Fig. 2.1. These fuzzy sets may have been chosen as reasonable representations of linguistic concepts of *very small, small, medium, large,* and *very large,* when applied to some variable v whose domain is the closed interval $[0, 100]$ (e.g., the utilization of a computing unit in a performance evaluation study, humidity, or the percentage of achieving a given goal).

For every $\alpha \in [0, 1]$, a given fuzzy set A yields the crisp set

$$^\alpha A = \{x \in X \mid A(x) \geq \alpha\}, \tag{2.8}$$

which is called an α-*cut* of A; it also yields the crisp set

$$^{\alpha+}A = \{x \in X \mid A(x) > \alpha\}, \tag{2.9}$$

FIGURE 2.1 Membership functions of five fuzzy sets that represent a variable whose values range from 0 to 100.

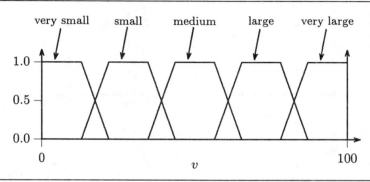

which is called a *strong α-cut* of A. Since $\alpha_1 < \alpha_2$ implies $^{\alpha_1}A \supseteq {}^{\alpha_2}A$ and $^{\alpha_1+}A \supseteq {}^{\alpha_2+}A$ the set of all distinct α-cuts as well as the set of all distinct strong α–cuts of any fuzzy set form nested sequences of crisp sets. The set ^{0+}A is called the *support* of A; the set ^1A is called the *core* of A. When $^1A \neq \emptyset$, A is called *normal*; otherwise it is called *subnormal*. The value

$$h(A) = \sup_{x \in X} A(x) \qquad (2.10)$$

is called the *height* of A and the value

$$p(A) = \inf_{x \in A} A(X) \qquad (2.11)$$

is called the *plimth* of A. The set

$$\Lambda(A) = \{\alpha \in [0,1] \mid A(x) = \alpha \text{ for some } x \in X\} \qquad (2.12)$$

is called the *level set* of A.

Given an arbitrary fuzzy set A, it is uniquely represented by the associated sequence of its α-cuts via the formula,

$$A(x) = \sup_{\alpha \in [0,1]} \alpha \cdot {}^{\alpha}A(x), \qquad (2.13a)$$

where $^{\alpha}A$ denotes the membership (characteristic) function of the α–cut of A and sup designates supremum (or the maximum, when X is finite). Alternately, A is uniquely represented by the strong α–cuts of A via the formula

$$A(x) = \sup_{\alpha \in [0,1]} \alpha \cdot {}^{\alpha+}A(x). \qquad (2.13b)$$

Equations (2.13a) and (2.13b), usually referred to as *decomposition theorems* of fuzzy sets, establish an important connection between fuzzy sets

FIGURE 2.2 From real numbers to fuzzy intervals: real number A; crisp interval B; fuzzy number C; fuzzy interval D

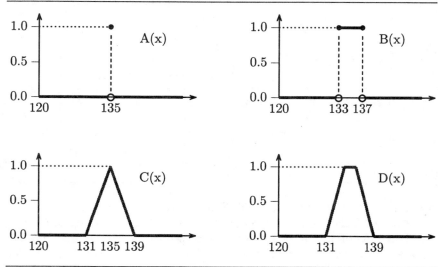

and crisp sets. This connection provides us with a criterion for generalizing properties of crisp sets into their fuzzy counterparts: when a fuzzy set satisfies a property that is claimed to be a generalization of a property established for crisp sets, this property should be preserved (in the crisp sense) in all α-cuts or strong α-cuts of the fuzzy set [Klir and Yuan, 1995a]. For example, all α-cuts of a convex fuzzy set should be convex crisp sets; the cardinality of a fuzzy set should yield for each α the cardinality of its α-cut; each α-cut of a properly defined fuzzy equivalence relation (or fuzzy compatibility relation, fuzzy ordering relation, etc.) should be an equivalence relation (or compatibility relation, ordering relation, etc., respectively) in the classical sense. Every property of a fuzzy set that is preserved in all its α-cuts is called a *cutworthy property*.

As illustrated by the example in Fig. 2.1, the notion of a fuzzy set allows us to generalize the concepts of a real number and an interval of real numbers to those of a fuzzy number and a fuzzy interval. Fuzzy intervals are, in general, normal and convex fuzzy sets defined on the set of real numbers whose α-cuts are closed intervals of real numbers and whose supports are bounded. *Fuzzy numbers* are special fuzzy intervals. A fuzzy number A is a fuzzy interval for which $A(x) = 1$ for exactly one $x \in X$. The distinction between fuzzy numbers and fuzzy intervals is often not made in the literature and both are called fuzzy numbers.

A triangular fuzzy number, $t\tilde{r}$, for example is named for its shape. Its membership function is given by two line segments, one segment rising from the point $\langle a, 0 \rangle$ to $\langle m, 1 \rangle$ and the second segment falling from $\langle m, 1 \rangle$

to $\langle b, 0 \rangle$. Its support is the closed interval of the real numbers $[a, b]$. A triangular fuzzy number can be specified by the ordered triple $\langle a, m, b \rangle$ with $a \leq m \leq b$. Clearly, the open interval (a, b) is the support of $\langle a, m, b \rangle$ and the singleton $\{m\}$ is its core. A trapezoidal fuzzy interval, $t\tilde{p}$, can be specified by an ordered quadruple $\langle a, l, r, b \rangle$ with $a \leq l \leq r \leq b$ and a membership function consisting of three line segments. The first segment rises from $\langle a, 0 \rangle$ to $\langle l, 1 \rangle$, the second segment is a horizontal line that has a constant value of one and that stretches from $\langle l, 1 \rangle$ to $\langle r, 1 \rangle$, and the third segment drops from $\langle r, 1 \rangle$ to $\langle b, 0 \rangle$. That is, the open interval (a, b) is the support of $\langle a, l, r, b \rangle$ and the closed interval $[l, r]$ is its core. For example, fuzzy set C in Fig. 2.2 is the triangular fuzzy number $\langle 131, 135, 139 \rangle$ and fuzzy set D in Fig. 2.2 is the trapezoidal number $\langle 131, 134, 136, 139 \rangle$. The generic membership function for a triangular fuzzy number $t\tilde{r}[a, m, b]$ is defined for each $x \in X$ by the formula

$$t\tilde{r}[a, m, b](x) = \begin{cases} \dfrac{x - a}{m - a} & \text{for } x \in [a, m] \\ \dfrac{x - b}{m - b} & \text{for } x \in (m, b] \\ 0 & \text{otherwise} \end{cases} \tag{2.14}$$

and for a trapezoidal fuzzy interval $t\tilde{p}[a, l, r, b]$ by the formula

$$t\tilde{p}[a, l, r, b](x) = \begin{cases} \dfrac{x - a}{l - a} & \text{for } x \in [a, l] \\ 1 & \text{for } x \in (l, r) \\ \dfrac{x - b}{r - b} & \text{for } x \in [r, b] \\ 0 & \text{otherwise} \end{cases} \tag{2.15}$$

An arithmetic of fuzzy numbers (and fuzzy intervals), which is a generalization of the well-established interval-valued arithmetic [Moore, 1966, 1979], is now rather well developed [Kaufmann and Gupta, 1985; Klir and Yuan, 1995a]. However, fuzzy arithmetic under various requisite constraints is still rather underdeveloped [Klir and Cooper, 1996].

A *fuzzy system* is a system in which states of the individual variables or any combination of them are fuzzy sets, usually fuzzy numbers or fuzzy intervals (as exemplified in Fig. 2.2). Viewing states of a system as fuzzy numbers or fuzzy intervals allows us to express the effect of *measurement errors* more faithfully than viewing them as real numbers or intervals of real numbers. Furthermore, fuzzy numbers, fuzzy intervals, and other types of fuzzy sets give us enough flexibility to represent, as closely as desirable, states characterized by *expressions in natural language* that are inherently vague. These expressions are, of course, strongly dependent on the context in which they are used. This implies that membership grade functions by which we attempt to capture the meaning of relevant linguistic expressions must be constructed in the context of each application. Various methods of knowledge acquisition developed in the area of knowledge engineering can be utilized in this construction. One method, which has lately been employed with great success in some applications of fuzzy systems, is based on training an appropriate neural network by data exemplifying the meaning of the linguistic expression involved in a particular application [Klir and Yuan, 1995a].

Two types of cardinalities are defined for fuzzy sets defined on a finite universal set X. One of them is called a *scalar cardinality* or, by some authors, a *sigma count;* it is a real number $|A|$ defined by the formula

$$|A| = \sum_{x \in X} A(x). \tag{2.16}$$

The other type, called a *fuzzy cardinality* and denoted by $card(A)$, is a fuzzy number defined for each $\alpha \in \Lambda(A)$ by the formula

$$card(A)(|^\alpha A|) = \alpha, \tag{2.17}$$

where $|^\alpha A|$ denotes the cardinality of the α-cut of A.

Fuzzy Operations

A single fuzzy set can be operated on by the application of a *fuzzy complement*. Several fuzzy sets can be combined by one of three types of aggregating operations: *fuzzy intersection*, *unions*, and *averaging operations*. Operations of each of these types are not unique. The whole scope of operations of each type can be conveniently captured by a class of functions distinguished from one another by distinct values of a parameter taken from a specific range of values. The choice of a particular operation is determined by the purpose for which it is used.

By far, the most important and common fuzzy complement, intersection and union operations are those defined by the formulas

$$\bar{A}(x) = 1 - A(x), \tag{2.18a}$$
$$(A \cap B)(x) = \min[A(x), B(x)], \tag{2.18b}$$
$$(A \cup B)(x) = \max[A(x), B(x)]. \tag{2.18c}$$

Axiomatic characterization of these operations, which are usually referred to as *standard fuzzy operations*, was investigated by Bellman and Gierz [1973]. The minimum operation is the only fuzzy intersection that is idempotent and cutworthy; similarly, the maximum operation is the only union that is idempotent and cutworthy. No fuzzy complement is cutworthy.

FUZZY COMPLEMENTS

An arbitrary complement operators, $co : [0, 1] \to [0, 1]$, must satisfy the following three axioms:

(co1) Membership dependency — The membership grade of x in the complement of A depends only on the membership grade of x in A.

(co2) Boundary condition — $co(0) = 1$ and $co(1) = 0$, that is co behaves as the ordinary complement for crisp sets.

(co3) Monotonicity — For all $a, b \in [0, 1]$, if $a < b$, then $co(a) \geq co(b)$, that is co is monotonic nonincreasing.

Two additional axioms, which are usually considered desirable, constrain the large family of functions that would be permitted by the above three axioms; they are:

(co4) Continuity — co is continuous.

(co5) Involution — co is involutive, that is $co(co(a)) = a$.

Some of the functions that conform to these five axioms besides the standard fuzzy complement are in the Sugeno class defined for all $a \in [0, 1]$ by

$$co[\lambda](a) = \frac{1 - a}{1 + \lambda a}, \tag{2.19}$$

with $\lambda \in (-1, \infty)$, and in the Yager class of fuzzy complements defined for all $a \in [0, 1]$ by

$$co[w](a) = (1 - a^w)^{1/w}, \tag{2.20}$$

with $w \in (0, \infty)$. Observe that the standard fuzzy complement, $\bar{A}(x) = 1 - A(x)$, is obtained as $co[\lambda = 0]$ or $co[w = 1]$.

An example of a fuzzy complement that conforms to (co1)–(co3) but not to (co4) and (co5) are the threshold fuzzy complements

$$co[t](a) = \left\{ \begin{array}{ll} 1 & \text{when } a \in [0, t] \\ 0 & \text{when } a \in (t, 1] \end{array} \right., \tag{2.21}$$

with $t \in [0, 1]$.

Subsequently, we shall write A^{co} for an arbitrary complement of the fuzzy set A; its membership function is $A^{co}(x) = co(A(x))$.

An equilibrium, e_{co}, of a fuzzy complement co, if it exists, is a number in $[0, 1]$ for which $co(e_{co}) = e_{co}$. Every fuzzy complement has at most one fuzzy equilibrium [Klir and Yuan, 1995a]. If a fuzzy complement is continuous (i.e., if it satisfies axioms (co1)–(co4)), the existence of its equilibrium is guaranteed [Klir and Yuan, 1995a]. For example, the equilibrium of fuzzy complements in the Yager class (2.20) are

$$e_{co}[w] = 0.5^{\frac{1}{w}} \tag{2.22}$$

for each $w \in (0, \infty)$.

FUZZY SET INTERSECTIONS

The intersection of two fuzzy sets must be a function that maps pairs of numbers in the unit interval into the unit interval $[0, 1]$, $i : [0, 1] \times [0, 1] \to [0, 1]$. It is now well established that *triangular norms* or *t–norms*, which have been extensively studied in the literature [Schweizer and Sklar, 1963] do possess all properties that are intuitively associated with fuzzy intersections. These functions are, for all $a, b, c, d \in [0, 1]$, characterized by he following axioms:

(i1) Boundary condition — $i(1, a) = a$.

(i2) Monotonicity — $i(a, b) \le i(c, d)$ whenever $a \le c$ and $b \le d$.

(i3) Commutativity — $i(a, b) = i(b, a)$.

(i4) Associativity — $i(a, i(b, c)) = i(i(a, b), c)$.

The largest t-norm is the minimum function and the smallest is

$$i_{min}(a, b) = \begin{cases} a & \text{when } b = 1 \\ b & \text{when } a = 1 \\ 0 & \text{otherwise} \end{cases}, \tag{2.23}$$

in the sense that if i is any t-norm then for any $a, b \in [0, 1]$

$$i_{min}(a, b) \le i(a, b) \le \min(a, b). \tag{2.24}$$

A host of t-norms have been proposed to deal with specific problems. A selection of some well-known parametric classes of t-norms are given in Table 2.1. Various procedures are now available for obtaining these and other classes of t–norms [Klir and Yuan, 1995a].

FUZZY SET UNIONS

The union of two fuzzy sets must be a function that maps pairs of numbers in the unit interval into the unit interval, $u : [0, 1] \times [0, 1] \to [0, 1]$. As is well known, functions known as *triangular conorms* or *t–conorms*, possess all the properties that are intuitively associated with fuzzy unions. They are characterized for all $a, b, c, d \in [0, 1]$ by the following axioms:

(u1) Boundary condition — $u(0, a) = a$.

(u2) Monotonicity — $u(a, b) \le u(c, d)$ whenever $a \le c$ and $b \le d$.

(u3) Commutativity — $u(a, b) = u(b, a)$.

(u4) Associativity — $u(a, u(b, c)) = u(u(a, b), c)$.

TABLE 2.1 Some classes of t–norms

FORMULA	PARAMETER	ORIGINATOR / YEAR
$i_{\min}$		
$[\max(0, a^p + b^p - 1)]^{\frac{1}{p}}$	$p \neq 0$	Schweizer&Sklar / 1983
$\dfrac{ab}{\gamma + (1 - \gamma)(a + b - ab)}$	$\gamma \in (0, \infty)$	Hamacher / 1978
$\log_s \left[1 + \dfrac{(s^a - 1)(s^b - 1)}{s - 1} \right]$	$s \in (0, \infty)$, $s \neq 1$	Frank / 1979
$1 - \min \left[1, ((1 - a)^w + (1 - a)^w)^{\frac{1}{w}} \right]$	$w \in (0, \infty)$	Yager / 1980a
$\dfrac{ab}{\max(a, b, \alpha)}$	$\alpha \in [0, 1]$	Dubois&Prade / 1980b
$\left[1 + \left[(\frac{1}{a} - 1)^\lambda + (\frac{1}{b} - 1)^\lambda \right]^{\frac{1}{\lambda}} \right]^{-1}$	$\lambda \in (0, \infty)$	Dombi / 1982

The smallest t-conorm is the maximum function and the largest is

$$u_{\max}(a, b) = \begin{cases} a & \text{when } b = 0 \\ b & \text{when } a = 0 \\ 1 & \text{otherwise} \end{cases}, \qquad (2.25)$$

in the sense that if u is any t-conorm then for any $a, b \in [0, 1]$

$$\max(a, b) \leq u(a, b) \leq u_{\max}(a, b). \qquad (2.26)$$

A host of t-conorms have been proposed to deal with specific problems. Some well known parameterized classes of t–conorms are given in Table 2.2.

AGGREGATION OPERATIONS

In addition to fuzzy intersections and fuzzy unions, fuzzy membership values may also be aggregated by *averaging operations*. These operations have no counterparts in classical set theory. Since averaging operations are in

TABLE 2.2 Some classes of t-conorms

FORMULA	PARAMETER	ORIGINATOR YEAR
$u_{\max}$		
$1 - [\max(0, (1-a)^p + (1-b)^p - 1)]^{\frac{1}{p}}$	$p \neq 0$	Schweizer&Sklar 1983
$\dfrac{a + b - (\gamma - 2)ab}{1 + (\gamma - 1)ab}$	$\gamma \in (0, \infty)$	Hamacher 1978
$1 - \log_s \left[1 + \dfrac{\left(s^{1-a} - 1\right)\left(s^{1-b} - 1\right)}{s - 1} \right]$	$s \in (0, \infty)$, $s \neq 1$	Frank 1979
$\min \left[1, (a^w + b^w)^{\frac{1}{w}} \right]$	$w \in (0, \infty)$	Yager 1980a
$1 - \dfrac{(1-a)(1-b)}{\max(a, b, 1 - \alpha)}$	$\alpha \in [0, 1]$	Dubois&Prade 1980b
$\left[1 + \left[\left(\frac{1}{a} - 1\right)^{-\lambda} + \left(\frac{1}{b} - 1\right)^{-\lambda} \right]^{-\frac{1}{\lambda}} \right]^{-1}$	$\lambda \in (0, \infty)$	Dombi 1982

general not associative, they must be defined as functions of m arguments ($m \geq 2$). That is, an averaging operation h is a function of the form

$$h : [0, 1]^m \to [0, 1]. \tag{2.27}$$

Averaging operations are characterized by the following set of axioms:

(h1) Idempotency — for all $a \in [0, 1]$,

$$h(a, a, a, \dots, a) = a. \tag{2.28}$$

(h2) Monotonicity — for any pair of m–tuples of real numbers in $[0, 1]$, $\langle a_1, a_2, a_3, \dots, a_m \rangle$ and $\langle b_1, b_2, b_3, \dots, b_m \rangle$, if $a_k \leq b_k$ for all $k \in \mathbb{N}_m$ then

$$h(a_1, a_2, a_3, \dots, a_m) \leq h(b_1, b_2, b_3, \dots, b_m). \tag{2.29}$$

It is significant that any function h that satisfies these axioms produces numbers that, for any m–tuple $\langle a_1, a_2, a_3, \dots, a_m \rangle \in [0, 1]^m$, lie in the closed interval defined by the inequalities

$$\min(a_1, a_2, a_3, \dots, a_m) \leq h(a_1, a_2, a_3, \dots, a_m) \leq \max(a_1, a_2, a_3, \dots, a_m). \tag{2.30}$$

The min and max operations qualify, due to their idempotency, not only as fuzzy counterparts of classical set intersection and union, respectively, but also as extreme averaging operations.

An example of a class of symmetric averaging operations are *generalized means*, which are defined for all m–tuples $\langle a_1, a_2, a_3, \ldots, a_m \rangle$ in $[0,1]^m$ by the formula

$$h_p(a_1, a_2, a_3, \ldots, a_m) = \frac{1}{m}(a_1^p + a_2^p + a_3^p + \ldots + a_m^p)^{\frac{1}{p}}, \qquad (2.31)$$

where p is a parameter whose range is the set of all real numbers excluding 0. For $p = 0$, h_p is not defined; however for $p \to 0$, h_p converges to the well known geometric mean. That is, we take

$$h_0(a_1, a_2, a_3, \ldots, a_m) = (a_1 a_2 a_3 \ldots a_m)^{\frac{1}{m}}. \qquad (2.32)$$

For $p \to -\infty$ and $p \to \infty$, h_p converges to the min and max operations, respectively.

Fuzzy Subsethood

Given two fuzzy sets A and B, we say that A is a *subset* of B if and only if

$$A(x) \le B(x) \qquad (2.33)$$

for all $x \in X$. If neither of the sets is a subset of the other, it is useful to express the extent to which either is a subset of the other one. Consider the *extent of subsethood* of A in B, for example, as it is expressed by the function

$$sub(A, B) = \frac{|A \cap B|}{|A|}, \qquad (2.34)$$

where $\cap$ denotes the standard fuzzy intersection (minimum operator) and $|A \cap B|$, $|A|$ are scalar cardinalities of the respective fuzzy sets. Observe that

$$0 \le sub(A, B) \le 1. \qquad (2.35)$$

The minimum, $sub(A, B) = 0$, is obtained when sets A and B do not overlap, while the maximum, $sub(A, B) = 1$ is obtained when A is a subset of B.

Cylindric Extensions

A fuzzy set A_Z defined on a universal set $Z = X \times Y$ is called a *joint fuzzy set* or, more commonly, a *fuzzy relation*. The fuzzy sets induced by the fuzzy relation A_Z on the two universes X and Y via the formulas, $A_X(x) = \max_{y \in Y}[A(x, y)]$ and $A_Y(y) = \max_{x \in X}[A(x, y)]$ are called *marginal fuzzy sets* or *projections* of A_Z. If A_X is a fuzzy set defined upon X then the

cylindric extension of A_X to $Z = X \times Y$ is the fuzzy relation defined
by $A_{X \uparrow Z}(x, y) = A_X(x)$ for all $y \in Y$. Similarly, for a fuzzy set A_Y, the
cylindric extension to $Z = X \times Y$ is defined by the membership function
$A_{Y \uparrow Z}(x, y) = A_Y(y)$ for all $x \in X$. If we have two fuzzy sets A_X and
A_Y then the *cylindric closure* of these fuzzy sets is a fuzzy set defined on
$Z = X \times Y$ whose membership function is given by the formula $A_{\hat{Z}}(x, y) =$
$\min[A(x), A(y)]$. For simplicity we denote this cylindric closure by $\hat{Z}$ in
the subsequent portions of this book. Notice that the cylindric closure of
A_X and A_Y is the standard fuzzy intersection of the individual cylindric
extensions of A_X and A_Y. It is the largest fuzzy relation on $X \times Y$ whose
projections are fuzzy sets A_X and A_Y

It should be obvious that the cylindric closure of projected marginal
fuzzy sets is not in general equal to the original fuzzy relation. For example
let $Z = X \times Y = \{a, b\} \times \{1, 2\}$ with $A_Z(a, 1) = 1.0$, $A_Z(a, 2) = 0.7$,
$A_Z(b, 1) = 0.6$, and $A_Z(b, 2) = 0.2$. By the definition of marginal fuzzy
sets we get as the projection of A_Z into X the marginal fuzzy set A_X
with $A_X(a) = 1.0$ and $A_X(b) = 0.6$. The projection of A_Z into Y is the
marginal fuzzy set A_Y with $A_Y(1) = 1.0$ and $A_Y(2) = 0.7$. The cylindric
closure of the two marginal fuzzy sets A_X and A_Y is the fuzzy set $A_{\hat{Z}}$ with
$A_{\hat{Z}}(a, 1) = 1.0$, $A_{\hat{Z}}(a, 2) = 0.7$, $A_{\hat{Z}}(b, 1) = 0.6$, and $A_{\hat{Z}}(b, 2) = 0.6$.

A fuzzy relation A_Z is said to be noninteractive if and only if

$$A_Z(x, y) = \min[A_X(x), A_Y(y)]. \tag{2.36}$$

with $Z = X \times Y$. Another way to say this is that $A_Z = A_{\hat{Z}}$; that is,
a fuzzy relation is noninteractive if it is equal to the cylindric closure of
its marginals. The definitions of this section can easily be extended to n-
dimensional fuzzy relations.

Types of Fuzzy Sets

For some applications, it is useful to define fuzzy sets in terms of more
general forms of membership grade functions. An important form is

$$A : X \to L, \tag{2.37}$$

where L denotes a lattice. Fuzzy sets defined by functions of this form are
called *L-fuzzy sets*. Lattice L may, for example, consist of a class of closed
intervals in $[0, 1]$. Membership degrees are in this case defined imprecisely,
by closed subintervals of $[0, 1]$. Fuzzy sets with this property are called
interval–valued fuzzy sets. When L is a class of fuzzy numbers defined on
$[0, 1]$, we obtain *fuzzy sets of type-2*.

These more general fuzzy sets as well as other aspects of fuzzy set theory
are not relevant to our discussion of uncertainty measures and principles
and, hence, there is no need to cover them here. A more comprehensive
coverage of fuzzy set theory can be found in several books [Dubois and

Prade, 1980a; Kandel, 1986; Klir and Yuan, 1995a; Nguyen and Walker, 1997; Novak, 1989; Zimmermann, 1985]. The development of key ideas in fuzzy set theory can be best traced through the papers of Zadeh [Klir and Yuan, 1996; Yager, et al., 1987]

2.3   Fuzzy Measure Theory

Fuzzy measure theory must be clearly distinguished from fuzzy set theory. While the latter is an outgrowth of classical set theory, the former is an outgrowth of classical measure theory.

The two theories may be viewed as complementary in the following sense. In fuzzy set theory, all objects of interest are precise and crisp; the issue is how much each given object is compatible with the concept represented by a given fuzzy set. In fuzzy measure theory, all considered sets are crisp, and the issue is the likelihood of membership in each of these sets of an object whose characterization is imprecise and, possibly, fuzzy. That is, while uncertainty in fuzzy set theory is associated with boundaries of sets, uncertainty in fuzzy measure theory is associated with boundaries of objects.

Given a universal set X and a non-empty family $\mathcal{C}$ of subsets of X (usually with an appropriate algebraic structure), a *fuzzy measure* (or *regular nonadditive measure*), g, on $\langle X, \mathcal{C} \rangle$ is a function

$$g : \mathcal{C} \to [0, 1] \tag{2.38}$$

that satisfies the following requirements:

(g1) Boundary conditions — $g(\emptyset) = 0$ when $\emptyset \in \mathcal{C}$ and $g(X) = 1$ when $X \in \mathcal{C}$.

(g2) Monotonicity — for all $A, B \in \mathcal{C}$, if $A \subseteq B$, then $g(A) \leq g(B)$.

(g3) Continuity from below — for any increasing sequence $A_1 \subseteq A_2 \subseteq A_3 \subseteq \ldots$ of sets in $\mathcal{C}$, if $\bigcup_{i=1}^{\infty} A_i \in \mathcal{C}$ then

$$\lim_{i \to \infty} g(A_i) = g\left(\bigcup_{i=1}^{\infty} A_i \right). \tag{2.39}$$

(g4) Continuity from above — for any decreasing sequence $A_1 \supseteq A_2 \supseteq A_3 \supseteq \ldots$ of sets in $\mathcal{C}$, if $\bigcap_{i=1}^{\infty} A_i \in \mathcal{C}$ then

$$\lim_{i \to \infty} g(A_i) = g\left(\bigcap_{i=1}^{\infty} A_i \right). \tag{2.40}$$

A few remarks regarding this definition are needed. First, functions that satisfy requirements (g1), (g2) and only one of the requirements (g3) and (g4) are equally important in fuzzy set theory. If only (g3) is satisfied, the function is called a *lower semicontinuous fuzzy measure*; if only (g4) is satisfied, it is called an *upper semicontinuous fuzzy measure*. Secondly, when the universal set X is finite, requirements (g3) and (g4) are trivially satisfied and may thus be disregarded. Third, it is sometimes needed to define fuzzy measures in a more general way by extending the range of function g to the set of all nonnegative real numbers and excluding the second boundary condition $g(X) = 1$. This generalization is not applicable when fuzzy measures are utilized for characterizing uncertainty. Fourth, in this book, $\mathcal{C}$ is assumed to be a σ–algebra: $X \in \mathcal{C}$, and if $A, B \in \mathcal{C}$, then $A \cup B \in \mathcal{C}$ and $A - B \in \mathcal{C}$. In most cases, $\mathcal{C}$ is the power set, $\mathcal{P}(X)$, of X.

We can see that fuzzy measures, as defined here, are generalizations of probability measures [Billingsley, 1986] or, when conceived in the broader sense, they are generalizations of classical measures [Halmos, 1950]. In either case, the generalization is obtained by replacing the additivity requirement with the weaker requirements of monotonicity and continuity or, at least, semicontinuity. This generalization was first conceived by Sugeno [1974]. A comprehensive and up-to-date introduction to fuzzy measure theory is the subject of a graduate text by Wang and Klir [1992]. Various aspects of fuzzy measure theory are also covered in books by Denneberg [1994], Grabisch et al. [1995], and Pap [1995].

Our primary interest in this book does not involve the full scope of fuzzy measure theory, but only three of its branches: evidence theory, probability theory, and possibility theory. Relevant properties of these theories are introduced in the rest of this chapter. Fuzzy measure theory is covered here because it represents a broad, unifying framework for future research regarding uncertainty-based information.

One additional remark should be made. Fuzzy measure theory, as well as any of its branches, may be combined with fuzzy set theory. That is, function g characterizing a fuzzy measure may be defined on fuzzy sets rather than crisp sets [Wang and Klir, 1992].

2.4   Evidence Theory

Evidence theory is based on two dual semicontinuous nonadditive measures (fuzzy measures): belief measures and plausibility measures. Given a universal set X, assumed here to be finite, a *belief measure* is a function

$$\mathrm{Bel} : \mathcal{P}(X) \rightarrow [0, 1] \tag{2.41}$$

such that $\text{Bel}(\emptyset) = 0$, $\text{Bel}(X) = 1$, and

$$
\begin{aligned}
\text{Bel}(A_1 \cup A_2 \cup \cdots \cup A_n) \geq & \sum_j \text{Bel}(A_j) \\
& - \sum_{j<k} \text{Bel}(A_j \cap A_k) \\
& + \sum_{j<k<l} \text{Bel}(A_j \cap A_k \cap A_l) \\
& \vdots \\
& + (-1)^{n+1} \text{Bel}(A_1 \cap A_2 \cap \cdots \cap A_n)
\end{aligned} \tag{2.42}
$$

for all possible families of subsets of X. Due to the inequality (2.42), belief measures are called *monotone of order* ∞. This property of belief measures implies that they are *superadditive* in the sense that

$$
\text{Bel}(A \cup B) \geq \text{Bel}(A) + \text{Bel}(B) \tag{2.43}
$$

for any disjoint sets $A, B \in \mathcal{P}(X)$. When X is infinite, function Bel is also required to be continuous from above.

A plausibility measure is a function

$$
\text{Pl} : \mathcal{P}(X) \to [0, 1] \tag{2.44}
$$

such that $\text{Pl}(\emptyset) = 0$, $\text{Pl}(X) = 1$, and

$$
\begin{aligned}
\text{Pl}(A_1 \cap A_2 \cap \cdots \cap A_n) \leq & \sum_j \text{Pl}(A_j) \\
& - \sum_{j<k} \text{Pl}(A_j \cup A_k) \\
& + \sum_{j<k<l} \text{Pl}(A_j \cup A_k \cup A_l) \\
& \vdots \\
& + (-1)^{n+1} \text{Pl}(A_1 \cup A_2 \cup \cdots \cup A_n)
\end{aligned} \tag{2.45}
$$

for all possible families of subsets of X. Due to (2.45), plausibility measures are called *alternate of order* ∞. This property of plausibility measures implies that they are *subadditive* in the sense that

$$
\text{Pl}(A \cup B) \leq \text{Pl}(A) + \text{Pl}(B) \tag{2.46}
$$

for any disjoint sets $A, B \in \mathcal{P}(X)$. When X is infinite, function Pl is also required to be continuous from below.

It is well known that either of the two measures is uniquely determined from the other by the equation

$$\text{Pl}(A) = 1 - \text{Bel}(\bar{A}) \tag{2.47}$$

for all $A \in \mathcal{P}(X)$, where $\bar{A}$ is the crisp complement of A. It is also well known that

$$\text{Pl}(A) \geq \text{Bel}(A) \tag{2.48}$$

for each $A \in \mathcal{P}(X)$. In the special case of equality in (2.48) for all $A \in \mathcal{P}(X)$, we obtain a classical (additive) probability measure.

Evidence theory, which has become an important tool for dealing with uncertainty, is best covered in a book by Shafer [1976]. Its position in fuzzy measure theory is described by Wang and Klir [1992]. The theory, often referred to as *Dempster–Shafer theory*, is also covered in books by Guan and Bell [1991, 1992] and Kohlas and Monney [1995].

Belief and plausibility measures can conveniently be characterized by a function

$$m : \mathcal{P}(X) \to [0, 1], \tag{2.49}$$

which is required to satisfy two conditions:

$$(\text{i}) \quad m(\emptyset) = 0 \; ; \tag{2.50}$$

$$(\text{ii}) \quad \sum_{A \in \mathcal{P}(X)} m(A) = 1 \, .$$

This function is called a *basic probability assignment*. For each set $A \in \mathcal{P}(X)$, the value $m(A)$ expresses the proportion to which all available and relevant evidence supports the claim that a particular element of X, whose characterization in terms of relevant attributes is deficient, belongs to the set A. This value, $m(A)$, pertains solely to one set, set A; it does not imply any additional claims regarding subsets of A. If there is some additional evidence supporting the claim that the element belongs to a subset of A, say $B \subseteq A$, it must be expressed by another value $m(B)$.

Given a basic probability assignment m, the corresponding belief measure and plausibility measure are determined for all sets $A \in \mathcal{P}(X)$ by the formulas

$$\text{Bel}(A) \;\; = \sum_{B \mid B \subseteq A} m(B) \; , \tag{2.51}$$

$$\text{Pl}(A) \;\; = \sum_{B \mid B \cap A \neq \emptyset} m(B). \tag{2.52}$$

Inverse procedures are also possible. Given, for example, a belief measure Bel, the corresponding basic probability assignment m is determined for all

$A \in \mathcal{P}(X)$ by the formula

$$m(A) = \sum_{B|B \subseteq A} (-1)^{|A-B|} \text{Bel}(B), \qquad (2.53)$$

where $|A - B|$ is the cardinality of the set difference of A and B, as proven by Shafer [1976]. If a plausibility measure is given, it can be converted to the associated belief measure by Eq. (2.47), and Eq. (2.53) is then applicable to make a conversion to function m. Hence, each of the three function, m, Bel and Pl, is sufficient to determine the other two.

Given a basic probability assignment, every set $A \in \mathcal{P}(X)$ for which $m(A) \neq 0$ is called a *focal element*. The pair $\langle \mathcal{F}, m \rangle$, where $\mathcal{F}$ denotes the set of all focal elements induced by m is called a *body of evidence*.

Total ignorance is expressed in evidence theory by $m(X) = 1$ and $m(A) = 0$ for all $A \neq X$. Full certainty is expressed by $m(\{x\}) = 1$ for one particular element of x and $m(A) = 0$ for all $A \neq \{x\}$.

As an example, let $X = \{x_1, x_2, x_3\}$ and let

$$
\begin{aligned}
m(\{x_1, x_2\}) &= 0.3 \qquad (2.54) \\
m(\{x_3\}) &= 0.1 \\
m(\{x_2, x_3\}) &= 0.2 \\
m(X) &= 0.4 \, .
\end{aligned}
$$

be a given basic probability assignment on $\mathcal{P}(X)$. The focal set of this basic probability assignment is the set

$$\mathcal{F} = \{\{x_1, x_2\}, \{x_3\}, \{x_2, x_3\}, \{x_1, x_2, x_3\}\}; \qquad (2.55)$$

that is, we assume that $m(A) = 0$ for all $A \notin \mathcal{F}$.

Using the given basic probability assignment we can calculate the belief and plausibility of any subset of X. For example, our belief in $\{x_2, x_3\}$ is

$$
\begin{aligned}
\text{Bel}(\{x_2, x_3\}) &= m(\{x_2, x_3\}) + m(\{x_3\}) \qquad (2.56) \\
&= 0.2 + 0.1 \\
&= 0.3.
\end{aligned}
$$

since $\{x_2, x_3\}$ and $\{x_3\}$ are the only subsets of $\{x_2, x_3\}$ in the focal set. The plausibility of $\{x_3\}$ is

$$
\begin{aligned}
\text{Pl}(\{x_3\}) &= m(X) + m(\{x_2, x_3\}) + m(\{x_3\}) \qquad (2.57) \\
&= 0.4 + 0.2 + 0.1 \\
&= 0.7,
\end{aligned}
$$

TABLE 2.3 An example of a basic probability assignment and the associated belief and plausibility measures.

Set	m	Bel	Pl
$\emptyset$	0.0	0.0	0.0
$\{x_1\}$	0.0	0.0	0.7
$\{x_2\}$	0.0	0.0	0.9
$\{x_1, x_2\}$	0.3	0.3	0.9
$\{x_3\}$	0.1	0.1	0.7
$\{x_1, x_3\}$	0.0	0.1	1.0
$\{x_2, x_3\}$	0.2	0.3	1.0
X	0.4	1.0	1.0

since X, $\{x_2, x_3\}$, and $\{x_3\}$ are in the focal set and their intersection with $\{x_3\}$ is non-empty.

Table 2.3 is complete listing of the basic probability assignment, belief, and plausibility of all subsets of X for this example.

To investigate ways of measuring the amount of uncertainty represented by each body of evidence, it is essential to understand properties of bodies of evidence whose focal elements are subsets of the Cartesian product of two sets. That is, we need to examine basic probability assignments of the form

$$m : \mathcal{P}(X \times Y) \to [0, 1], \qquad (2.58)$$

where X and Y denote universal sets pertaining to two distinct domains of inquiry (e.g. two investigated variables), which may be connected in some fashion. Let m of this form be called a *joint basic probability assignment*. In this case, each focal element induced by m is a binary relation R on $X \times Y$. When R is projected on set X and on set Y, we obtain, respectively, the sets

$$R_X = \{x \in X \mid \langle x, y \rangle \in R \text{ for some } y \in Y\} \qquad (2.59)$$

and

$$R_Y = \{y \in Y \mid \langle x, y \rangle \in R \text{ for some } x \in X\}. \qquad (2.60)$$

These sets are instrumental in calculating marginal basic probability assignments m_X and m_Y from the given joint assignment m:

$$m_X(A) = \sum_{R \mid A = R_X} m(R) \text{ for all } A \in \mathcal{P}(X), \qquad (2.61)$$

and

$$m_Y(B) = \sum_{R|B=R_Y} m(R) \text{ for all } B \in \mathcal{P}(Y). \qquad (2.62)$$

Let the bodies of evidence associated with m_X and m_Y, $\langle \mathcal{F}_X, m_X \rangle$ and $\langle \mathcal{F}_Y, m_Y \rangle$, be called marginal bodies of evidence. These bodies are said to be noninteractive if and only if for all $A \in \mathcal{F}_X$ and all $B \in \mathcal{F}_Y$

$$m(A \times B) = m_X(A) \cdot m_Y(B) \qquad (2.63)$$

and

$$m(R) = 0 \text{ for all } R \neq A \times B. \qquad (2.64)$$

That is, two marginal bodies of evidence are noninteractive if and only if the only focal elements of the joint body of evidence are Cartesian products of focal elements of the marginal bodies and m is determined from m_X and m_Y by Eq. (2.63).

Example 2.1 *As an example, consider the body of evidence given in Table 2.4(a). Focal elements are subsets of the Cartesian product $X \times Y$, where $X = \{1,2,3\}$ and $Y = \{a,b,c\}$; they are defined in the table by their characteristic function. To emphasize that each focal element is, in fact, a binary relation on $X \times Y$, they are labeled $R_1, R_2, R_3, \ldots, R_{12}$. Employing Eqs. (2.61) and (2.62), we obtain the marginal bodies of evidence shown in Table 2.4(b). For example,*

$$
\begin{aligned}
m_X(\{2,3\}) &= m(R_1) + m(R_2) + m(R_3) = .25\,, \\
m_X(\{1,2\}) &= m(R_5) + m(R_8) + m(R_{11}) = .3\,, \\
m_Y(\{a\}) &= m(R_2) + m(R_7) + m(R_8) + m(R_9) = .25\,, \\
m_Y(\{a,b,c\}) &= m(R_3) + m(R_{10}) + m(R_{11}) + m(R_{12}) = .5\,,
\end{aligned}
$$

and similarly for the remaining sets A and B. We can easily verify that the joint basic assignment m is uniquely determined by the marginal basic assignments through Eq. (2.63). The marginal bodies of evidence are thus noninteractive. For example,

$$
\begin{aligned}
m(R_1) &= m_X(\{2,3\}) \cdot m_Y(\{b,c\}) \\
&= .25 \times .25 = .0625
\end{aligned}
$$

Observe that $\{2,3\} \times \{b,c\} = \{2b, 2c, 3b, 3c\} = R_1$. Similarly

$$
\begin{aligned}
m(R_{10}) &= m_X(\{1,3\}) \cdot m_Y(\{a,b,c\}) \\
&= .15 \times .5 = .075\,,
\end{aligned}
$$

where $R_{10} = \{1,3\} \times \{a,b,c\} = \{1a, 1b, 1c, 3a, 3b, 3c\}$.

TABLE 2.4 Joint and marginal bodies of evidence: illustration of independence(Example 2.1).

(a) Joint Body of evidence

	1a	1b	1c	2a	2b	2c	3a	3b	3c	$m(R_i)$
$R_1 =$	0	0	0	0	1	1	0	1	1	.0625
$R_2 =$	0	0	0	1	0	0	1	0	0	.0625
$R_3 =$	0	0	0	1	1	1	1	1	1	.125
$R_4 =$	0	1	1	0	0	0	0	1	1	.0375
$R_5 =$	0	1	1	0	1	1	0	0	0	.075
$R_6 =$	0	1	1	0	1	1	0	1	1	.075
$R_7 =$	1	0	0	0	0	0	1	0	0	.0375
$R_8 =$	1	0	0	1	0	0	0	0	0	.075
$R_9 =$	1	0	0	1	0	0	1	0	0	.075
$R_{10} =$	1	1	1	0	0	0	1	1	1	.075
$R_{11} =$	1	1	1	1	1	1	0	0	0	.15
$R_{12} =$	1	1	1	1	1	1	1	1	1	.15

The column groups above the 1a…3c headers are labelled $X \times Y$. Handwritten annotations: "X (123)" and "Y (abc)".

$$m : \mathcal{P}(X \times Y) \to [0,1]$$

(b) Marginal bodies of evidence

	1	2	3	$m_X(A)$
$A =$	0	1	1	.25
	1	0	1	.15
	1	1	0	.3
	1	1	0	.3

$$m_X : \mathcal{P}(X) \to [0,1]$$

	a	b	c	$m_Y(B)$
$B =$	0	1	1	.25
	1	0	0	.25
	1	1	1	.5

$$m_Y : \mathcal{P}(Y) \to [0,1]$$

Upper and Lower Probabilities

In Sec. 2.5, probability measures are viewed as a special case of the two measures employed in evidence theory. However the genesis of evidence theory was the work of Dempster [1967a,b] on upper and lower probabilities.

In probability theory, weights expressing evidential claims are assigned to individual elements of some universal set X. The probability of any subset of the sample space X is calculated by adding the weights of the subset's elements. Dempster examined the reverse problem. Suppose you know the probabilities of some subsets of a universal set. What can you say about the probabilities of the other subsets and of the singletons?

If the evidence is not contradictory then, usually, the problem does not allow for an exact solution. Instead, for each subset of the universe, a maximum and minimum probability consistent with the given probabilities can be calculated. The correct solution then lies somewhere in the interval between these two values.

The minimum consistent probability assigned to a set $A \subseteq X$ is called the lower probability of A and is denoted $P_*(A)$. The maximum consistent probability assigned to a set $A \subseteq X$ is called the upper probability of A and is denoted $P^*(A)$. The correct probability assigned to A, $Pro(A)$ (see Sec. 2.5), must be bracketed by these values, i.e.,

$$P_*(A) \leq Pro(A) \leq P^*(A) \tag{2.65}$$

or

$$Pro(A) \in [P_*(A), P^*(A)] \quad . \tag{2.66}$$

As expressed by Dempster, there is a family of probability measures, $\mathcal{P}$, bounded by the dual upper and lower probabilities and consistent with the given evidence,

$$\mathcal{P} = \{ Pro \mid P_*(A) \leq Pro(A) \leq P^*(A) \text{ for all } A \subseteq X \} \quad . \tag{2.67}$$

Let V be any variable defined upon X. That is, V is a function that maps X into the real numbers. Then, we can define an upper and lower probability distributions, F^* and F_*, on V by the formulas

$$F^*(v) = P^*(V < v), \tag{2.68}$$
$$F_*(v) = P_*(V < v), \tag{2.69}$$

where $v = V(x)$.

We can now define the upper and lower expected value of the random variable V [Dubois and Prade, 1985b] by the Lebesgue-Stieltjes integrals

$$E^*(V) = \int_{-\infty}^{+\infty} v \, dF^*(v), \tag{2.70}$$

$$E_*(V) = \int_{-\infty}^{+\infty} v \, dF_*(v), \tag{2.71}$$

respectively.

Of course $E_*(V) \leq E^*(V)$. The names "upper" and "lower" expectations are motivated by Dempster's identities

$$E^*(V) = \max_{P \in \mathcal{P}} E(V), \tag{2.72}$$

$$E_*(V) = \min_{P \in \mathcal{P}} E(V). \tag{2.73}$$

where $E(V)$ is the usual expected value of statistics.

In Shafer's reinterpretation of Dempster's work, the formulas for evidential expectation can be expressed in terms of belief and plausibility measures by the equations

$$\mathrm{Pl}(\{x_i|v_i \leq v\}) \;\;=\;\; F^*(v) = P^*(V < v), \tag{2.74}$$
$$\mathrm{Bel}(\{x_i|v_i \leq v\}) \;\;=\;\; F_*(V) = P_*(V < v). \tag{2.75}$$

This produces the expectations

$$E^*(V) \;\;=\;\; \int_{-\infty}^{+\infty} v d\mathrm{Pl}(\{x_i|v_i \leq v\}), \tag{2.76}$$

$$E_*(V) \;\;=\;\; \int_{-\infty}^{+\infty} v d\mathrm{Bel}(\{x_i|v_i \leq v\}), \tag{2.77}$$

where $v_i = V(x_i)$.

This level of coverage of evidence theory is sufficient for our purpose in this book. Before leaving this section, let us add a few historical and bibliographical remarks. In view of Dempster's work, evidence theory can be interpreted as a theory for dealing with imprecise probabilities: plausibility measures and belief measures play the roles of upper and lower probabilities. Evidence theory is also connected with the theory of random sets [Matheron, 1975]. In addition to the seminal book by Shafer [1976] and his useful overview [Shafer, 1990], many excellent papers on various aspects of evidence theory are now available; the following is a small sample: Smets [1988, 1990a,b], Strat [1990], Strat and Lowrance [1989], Yen [1989]. As shown by Yen [1990], evidence theory can be generalized to fuzzy sets.

2.5   Probability Theory

Consider a body of evidence $\langle \mathcal{F}, m \rangle$ in the sense of evidence theory. If the associated belief function is also additive, that is

$$\mathrm{Bel}(A \cup B) = \mathrm{Bel}(A) + \mathrm{Bel}(B) \quad \text{whenever} \quad A \cap B = \emptyset, \tag{2.78}$$

then we obtain a special type of belief measure — the classical probability measure. The following theorem [Shafer,1976] shows that whenever $\mathcal{F}$ consists solely of singletons, the associated belief and plausibility measures collapse into a single measure. The theorem also shows that the emerging single measure is a *probability measure* since it satisfies the condition of additivity.

Theorem 2.1 *A belief measure* Bel *on a finite power set* $\mathcal{P}(X)$ *is a probability measure if and only if its basic assignment* m *is given by* $m(\{x\}) = $ Bel$(\{x\})$ *and* $m(A) = 0$ *for all subsets* A *of* X *that are not singletons.*

Proof. Assume that Bel is a probability measure. For the empty set $\emptyset$, the theorem trivially holds, since $m(\emptyset) = 0$ by the definition of m. Let $A \neq \emptyset$ and assume $A = \{x_1, x_2, \ldots, x_n\}$. Then by repeated application of Eq. (2.78), we obtain

$$
\begin{aligned}
\text{Bel}(A) &= \text{Bel}(\{x_1\}) + \text{Bel}(\{x_2, x_3, \ldots, x_n\}) \\
&= \text{Bel}(\{x_1\}) + \text{Bel}(\{x_2\}) + \text{Bel}(\{x_3, x_4, \ldots, x_n\}) \\
&\;\;\vdots \\
&= \text{Bel}(\{x_1\}) + \text{Bel}(\{x_2\}) + \cdots + \text{Bel}(\{x_n\}) .
\end{aligned}
$$

Since Bel$(\{x\}) = m(\{x\})$ for any $x \in X$, by Eq. (2.51) we have

$$
\text{Bel}(A) = \sum_{i=1}^{n} m(\{x_i\}) .
$$

Hence, Bel is defined in terms of a basic probability assignment that focuses only on singletons.

Assume now that a basic probability assignment m is given such that

$$
\sum_{x \in X} m(\{x\}) = 1 .
$$

Then for any sets $A, B \in \mathcal{P}(X)$ such that $A \cap B = \emptyset$, we have

$$
\begin{aligned}
\text{Bel}(A) + \text{Bel}(B) &= \sum_{x \in A} m(\{x\}) + \sum_{x \in B} m(\{x\}) \\
&= \sum_{x \in A \cup B} m(\{x\}) = \text{Bel}(A \cup B)
\end{aligned}
$$

and, consequently, Bel is a probability measure. ∎

Any probability measure, *Pro*, on a finite set X can be uniquely determined by a *probability distribution function*

$$
p : X \rightarrow [0, 1] \tag{2.79}
$$

via the formula

$$
Pro(A) = \sum_{x \in A} p(x). \tag{2.80}
$$

From the standpoint of evidence theory, clearly

$$
p(x) = m(\{x\}) \tag{2.81}
$$

for all $x \in X$.

Within probability theory, restricted to probability measures, full certainty is expressed by $p(x) = 1$ for a particular $x \in X$, which conforms to the expression of full certainty in evidence theory. The expression of total ignorance in probability theory, on the other hand, which is given by

$$p(x) = \frac{1}{|X|} = m(\{x\}) \tag{2.82}$$

for all $x \in X$, is radically different from its counterpart in evidence theory (Sec. 2.4).

When a probability distribution function p is defined on a Cartesian product $X \times Y$, it is called a *joint probability distribution*. The associated *marginal probability distributions* are determined by the formulas

$$p_X(x) = \sum_{y \in Y} p(x, y) \tag{2.83}$$

for each $x \in X$ and

$$p_Y(y) = \sum_{x \in X} p(x, y) \tag{2.84}$$

for each $y \in Y$. The *noninteraction* of the marginal bodies of evidence is defined by the condition

$$p(x, y) = p_X(x) \cdot p_Y(y) \tag{2.85}$$

for all $x \in X$ and all $y \in Y$. Observe that Eq. (2.85) is a special case of Eq. (2.63), which expresses noninteraction in evidence theory.

2.6   Possibility Theory

We say that a family of subsets of a universal set is nested if these subsets can be ordered in such a way that each is contained in the next. For example, $A_1 \subset A_2 \subset A_3 \subset A_4 \subset X$ is a nested family of four subsets of a universal set X. In the literature, nested families of subsets are also called *chains*.

When it is required that the focal elements in a body of evidence $\langle \mathcal{F}, m \rangle$ be nested, the associated belief and plausibility measures are called *consonant*. This name appropriately reflects the fact that the degrees of evidence allocated to the focal elements that are nested have minimal conflict with each other; that is, nested focal elements are almost free of dissonance in evidence.

The requirement that focal elements be nested restricts belief and plausibility measures in a way that is characterized by the following theorem.

Theorem 2.2 *Given a consonant body of evidence $\langle \mathcal{F}, m \rangle$, the associated consonant belief and plausibility measures possess the following properties:*
   (i) $\text{Bel}(A \cap B) = \min[\text{Bel}(A), \text{Bel}(B)]$ *for all* $A, B \in \mathcal{P}(X)$;
   (ii) $\text{Pl}(A \cup B) = \max[\text{Pl}(A), \text{Pl}(B)]$ *for all* $A, B \in \mathcal{P}(X)$.

Proof. (i) Since the focal elements in $\mathcal{F}$ are nested, they may be linearly ordered by the subset relation. Let $\mathcal{F} = \{A_1, A_2, \ldots, A_n\}$ and assume that $A_i \subset A_j$ whenever $i < j$. Consider now arbitrary subsets A and B of X. Let i_1 be the largest integer i such that $A_i \subset A$ and let i_2 be the largest integer i such that $A_i \subset B$. Then $A_i \subset A$ and $A_i \subset B$ if and only if $i \leq i_1$ and $i \leq i_2$, respectively. Moreover, $A_i \subset A \cap B$ if and only if $i \leq \min[i_1, i_2]$. Hence,

$$
\begin{aligned}
\text{Bel}(A \cap B) &= \sum_{i=1}^{\min[i_1, i_2]} m(A_i) \\
&= \min \left[\sum_{i=1}^{i_1} m(A_i), \sum_{i=1}^{i_2} m(A_i) \right] \\
&= \min[\text{Bel}(A), \text{Bel}(B)].
\end{aligned}
$$

(ii) Assume that (i) holds. Then by Eq. (2.47),

$Bel(A) = 1 - Pl(\overline{A})$

$$
\begin{aligned}
\text{Pl}(A \cup B) &= 1 - \text{Bel}(\overline{A \cup B}) \\
&= 1 - \text{Bel}(\overline{A} \cap \overline{B}) \\
&= 1 - \min[\text{Bel}(\overline{A}), \text{Bel}(\overline{B})] \\
&= \max[1 - \text{Bel}(\overline{A}), 1 - \text{Bel}(\overline{B})] \\
&= \max[\text{Pl}(A), \text{Pl}(B)].
\end{aligned}
$$

for all $A, B \in \mathcal{P}(X)$.   ■

The special branch of evidence theory that deals only with bodies of evidence whose focal elements are nested is referred to as *possibility theory* [Dubois and Prade, 1988]. Special counterparts of belief measures and plausibility measures in possibility theory are called *necessity measures* and *possibility measures*, respectively.

Theorem 2.3 *Every possibility measure Pos on $\mathcal{P}(X)$ can be uniquely determined by a possibility distribution function*

$$
r : X \to [0, 1]
$$

via the formula

$$
Pos(A) = \max_{x \in A} r(x) \tag{2.86}
$$

for each $A \in \mathcal{P}(X)$.

Proof. We prove the theorem by induction on the cardinality of the set A. Let $|A| = 1$. Then, $A = \{x\}$, where $x \in X$, and Eq. (2.86) is trivially satisfied. Assume now that Eq. (2.86) is satisfied when $|A| = n - 1$, and let $A = \{x_1, x_2, \ldots, x_n\}$. Then by Theorem 2.2,

$$
\begin{aligned}
Pos(A) &= \max\left[Pos(\{x_1, x_2, \ldots, x_{n-1}\}), Pos(\{x_n\})\right] \\
&= \max\left[\max\left[Pos(\{x_1\}), Pos(\{x_2\}), \ldots, Pos(\{x_{n-1}\})\right], Pos(\{x_n\})\right] \\
&= \max\left[Pos(\{x_1\}), Pos(\{x_2\}), \ldots, Pos(\{x_n\})\right] \\
&= \max_{x \in A} r(x),
\end{aligned}
$$

so that Eq. (2.86) is satisfied for all $n \in \mathbb{N}$. ∎

The corresponding necessity measure, Nec, is then determined for all $A \in \mathcal{P}(X)$ by the formula

$$Nec(A) = 1 - Pos(\bar{A}) \tag{2.87}$$

which is a possibilistic counterpart of Eq. (2.47).

For all $A, B \in \mathcal{P}(X)$, possibility measures and necessity measures satisfy the equations

$$
\begin{aligned}
Pos(A \cup B) &= \max[Pos(A), Pos(B)], & (2.88) \\
Nec(A \cap B) &= \min[Nec(A), Nec(B)], & (2.89)
\end{aligned}
$$

and the inequalities

$$
\begin{aligned}
Pos(A \cap B) &\leq \min[Pos(A), Pos(B)], & (2.90) \\
Nec(A \cup B) &\geq \max[Nec(A), Nec(B)], & (2.91)
\end{aligned}
$$

Furthermore,

$$Nec(A) \leq Pos(A), \tag{2.92}$$

$$Pos(A) + Pos(\bar{A}) \geq 1, \tag{2.93}$$

$$Nec(A) + Nec(\bar{A}) \leq 1, \tag{2.94}$$

$$\max[Pos(A), Pos(\bar{A})] = 1, \tag{2.95}$$

$$\min[Nec(A), Nec(\bar{A})] = 0, \tag{2.96}$$

$$Pos(A) < 1 \implies Nec(A) = 0, \tag{2.97}$$

$$Nec(A) > 0 \implies Pos(A) = 1. \tag{2.98}$$

Let us introduce now a convenient notation, applicable to finite universal sets, in terms of which the connection between possibility distributions and basic probability assignments can be expressed by simple equations.

Assume the finite universe $X = \{x_1, x_2, \ldots, x_n\}$ and let $A_1 \subset A_2 \subset \ldots \subset A_n$, where $A_i = \{x_1, x_2, \ldots, x_i\}$ for $i \in \mathbb{N}_n$, be a complete sequence of nested subsets that contains all focal elements of a possibility measure Pos. That is, if $m(A) \neq 0$ then $A \in \{A_1, A_2,..., A_n\}$. Let $m_i = m(A_i)$ and $r_i = r(x_i)$ for all $i \in \mathbb{N}_n$. Then, the n-tuples

$$\mathbf{m} = \langle m_1, m_2, \ldots, m_n \rangle \tag{2.99a}$$

$$\mathbf{r} = \langle r_1, r_2, \ldots, r_n \rangle \tag{2.99b}$$

fully characterize the basic probability assignment and the possibility distribution, respectively, by which the possibility measure Pos is defined. The nested structure implies that $r_i \geq r_{i+1}$ for all $i \in \mathbb{N}_{n-1}$. That is, possibility distributions are in this formulation always ordered and $r_1 = 1$ in each possibility distribution. It is easy to show that

$$r_i = \sum_{k=i}^{n} m_k, \tag{2.99c}$$

$$m_i = r_i - r_{i+1} \tag{2.99d}$$

for all $i \in \mathbb{N}_n$, where $r_{n+1} = 0$ by convention [Klir and Folger, 1988].

Ordered possibility distributions of the same length can be partially ordered in the following way: given two possibility distributions

$$^i\mathbf{r} = \langle {}^ir_1, {}^ir_2, \ldots, {}^ir_n \rangle \tag{2.100}$$

and

$$^j\mathbf{r} = \langle {}^jr_1, {}^jr_2, \ldots, {}^jr_n \rangle \tag{2.101}$$

we define

$$^i\mathbf{r} \leq {}^j\mathbf{r} \text{ if and only if } {}^ir_k \leq {}^jr_k \tag{2.102}$$

for all $k \in \mathbb{N}_n$. This partial ordering forms a lattice whose join, $\vee$, and meet, $\wedge$, are defined, respectively, as

$$^i\mathbf{r} \vee {}^j\mathbf{r} = \langle \max[{}^ir_1, {}^jr_1], \max[{}^ir_2, {}^jr_2], \ldots, \max[{}^ir_n, {}^jr_n] \rangle \tag{2.103}$$

and

$$^i\mathbf{r} \wedge {}^j\mathbf{r} = \langle \min[{}^ir_1, {}^jr_1], \min[{}^ir_2, {}^jr_2], \ldots, \min[{}^ir_n, {}^jr_n] \rangle \tag{2.104}$$

for all pairs of possibility distributions of the same length n. The smallest possibility distribution is $\langle 1, 0, ..., 0 \rangle$, the greatest one is $\langle 1, 1, ..., 1 \rangle$; their counterparts expressing the basic probability assignment are $\langle 1, 0, ..., 0 \rangle$ and $\langle 0, 0, ..., 0, 1 \rangle$, respectively. We can see from this that in possibility theory the expression of full certainty and, contrary to probability theory, also the expression of total ignorance are exactly the same as in evidence theory.

In terms of fuzzy measure theory, a possibility measure Pos is a lower semicontinuous fuzzy measure for which

$$Pos(\emptyset) = 0, \tag{2.105}$$
$$Pos(X) = 1, \tag{2.106}$$

and

$$Pos\left(\bigcup_{i \in I} A_i\right) = \sup_{i \in I} Pos(A_i) \tag{2.107}$$

for any family $\{A_i \mid A_i \in \mathcal{P}(X), \ i \in I\}$ of subsets of X, where I is an arbitrary index set. This last property may also be expressed in an alternative way by the equation

$$Pos(A) = \sup_{x \in A} r(x) \tag{2.108}$$

for all $A \in \mathcal{P}(X)$, where

$$r(x) = Pos(\{x\}) \tag{2.109}$$

for all $x \in X$. A necessity measure Nec is a upper semicontinuous fuzzy measure for which

$$Nec(\emptyset) = 0, \tag{2.110}$$
$$Nec(X) = 1, \tag{2.111}$$

and

$$Nec\left(\bigcap_{i \in I} A_i\right) = \inf_{i \in I} Nec(A_i) \tag{2.112}$$

for any family of subsets of X defined by an arbitrary index set I.

It is often desirable that possibility theory be viewed as a separate formal mathematical theory. Either Eqs. (2.105)–(2.107), where Eq. (2.107) can be replaced with Eq. (2.108), or Eqs. (2.110)–(2.112) are usually employed as axioms of possibility theory. These two approaches are connected with one another by Eq. (2.87). The role of possibility theory in evidence theory is but one of its interpretations. Another, and perhaps the most visible interpretation of possibility theory, is connected with fuzzy sets. This interpretation was introduced by Zadeh [1978].

To explain the fuzzy set interpretation of possibility theory, let $\mathcal{X}$ denote a variable that takes values in a given set X, and let information about the actual value of the variable be expressed by a fuzzy proposition "$\mathcal{X}$ is F", where F is a standard fuzzy subset of X (i.e., $F(x) \in [0, 1]$ for all $x \in X$). To express information in measure-theoretic terms, it is natural to interpret the membership degree $F(x)$ for each $x \in X$ as the degree of possibility that $\mathcal{X} = x$. This interpretation induces a possibility distribution, r_F, on X that is defined by the equation

$$r_F(x) = F(x) \tag{2.113}$$

for all $x \in X$.

When the fuzzy set F in Eq. (2.113) is normal, the standard formulation of possibility theory, as outlined in this section, is directly applicable. However, the requirement that F be always a normal fuzzy subset of X is overly restrictive and cannot be guaranteed in some applications. How, then, to define the possibility measure and its dual necessity measure for a fuzzy proposition based on a fuzzy set that is not normal?

The answer to this question is not obvious. When Zadeh introduced possibility measures based on fuzzy sets (not necessarily normal), he did not address the issue of the dual necessity measures. To discuss this issue, let $h(F)$ denote the height of the given fuzzy set F and let $s(F)$ denote the support of F; furthermore, let

$$\bar{s}(F) = \{x \in X \mid F(x) = 0\} \tag{2.114}$$

denote the complement of the support of F.

When $h(F) < 1$, the interpretation expressed by Eq. (2.113) is not directly applicable. This can be seen, for example, from these facts: since

$$Pos_F\left(s(F)\right) = h(F) \quad \text{and} \quad Pos_F\left(\bar{s}(F)\right) = 0, \tag{2.115}$$

we obtain by Eq. (2.87) that $Nec_F\left(s(F)\right) = 1$. Hence,

$$Nec_F\left(s(F)\right) > Pos_F\left(s(F)\right), \tag{2.116}$$

which violates Eq. (2.92) and is not acceptable for the modalities of necessity and possibility. Trying to "fix" this problem by normalizing F (dividing $F(x)$ by $h(F)$ for each $x \in X$) is also not acceptable because it involves a distortion of the information supported by given evidence.

Following Zadeh [1979], Yager [1986] attempted to solve this problem by introducing a new function, $Cert$, defined for all $A \in \mathcal{P}(X)$ by the equation

$$Cert_F(A) = \min\left[Pos_F(A), Nec_F(A)\right], \tag{2.117}$$

and suggesting the use of this function (called a measure of certainty) instead of the function Nec. Although this new function satisfies the essential inequality of Eq. (2.92) for all fuzzy subsets of X, it has other severe drawbacks, as observed by Dubois and Prade [1987a]. Not only is its interpretation unclear, but above all, it does not satisfy Eq. (2.89), which is an essential property of necessity measures.

Dubois and Prade [1987a] suggested to resolve the problem by keeping the function Nec_F, but replacing Eq. (2.87) with the equation

$$Nec_F(A) = h(F) - Pos_F(\bar{A}). \tag{2.118}$$

Clearly Eq. (2.118) reduces to Eq. (2.87) when F is normal. Moreover, function Nec_F, defined by Eq. (2.118) satisfies both Eqs. (2.89) and (2.92) for every fuzzy subset F of X. Hence, this suggestion is more reasonable. However, it still has some drawbacks. First, it does not guarantee that $Pos_F(X) = Nec_F(X) = 1$. This is not a desirable result since X is always assumed to be chosen in such a way that $\mathcal{X}$ can only take values in X and, hence, its unknown value must be in X regardless of the available information about this value. Second, the suggested connection between Nec_F and Pos_F violates the properties described by Eqs. (2.97) and (2.98), which are fundamental to possibility theory. Third, a nonzero value is allocated to the empty set by the basic probability assignment function m_F associated with F; specifically,

$$m_F(\emptyset) = 1 - h(F),\qquad(2.119)$$

but it is required that $Nec_F(\emptyset) = 0$.

In the face of the mentioned drawbacks of previous proposals to modify the standard possibility theory in order to accommodate its connection with subnormal fuzzy sets, Klir and Harmanec [1995] proposed to deal with the issue by requiring that Eq. (2.106) be satisfied for all $F \in \hat{\mathcal{P}}(X)$. Unfortunately, their formulation, based on this requirement, is somewhat ambiguous. Let us present here an alternative formulation to eliminate the ambiguity.

To satisfy the requirement that $Pos_F(X) = 1$ when $h(F) < 1$, we need to revise the fuzzy set interpretation of possibility theory defined by Eq. (2.113). That is, r_F must be defined in such a way that

$$\sup_{x \in X} r_F(x) = 1.\qquad(2.120)$$

Since there is no reason to treat distinct elements of the universal set differently, the only sensible way to satisfy Eq. (2.120) is to increase the values of $r_F(x)$ equally for all elements $x \in X$ by the amount $1 - h(F)$. This means that the fuzzy set interpretation is expressed for all $x \in X$ by the equation

$$r_F(x) = F(x) + 1 - h(F).\qquad(2.121)$$

This is a generalized counterpart of Eq. (2.113) which is applicable to all fuzzy sets, regardless whether they are normal or not. For normal fuzzy sets, clearly, (2.121) collapses to (2.113).

It is easy to verify that all the properties of possibility theory are preserved under this interpretation. The following are some interesting values of Pos_F and Nec_F:

$$Pos_F(s(F)) = 1,\qquad(2.122)$$

$$Pos_F(\bar{s}(F)) = \begin{cases} 1 - h(F) & \text{when } \bar{s}(F) \neq \emptyset \\ 0 & \text{when } \bar{s}(F) = \emptyset \end{cases},\qquad(2.123)$$

$$Nec_F(s(F)) = 1 - Pos_F(\bar{s}(F)),\qquad(2.124)$$

$$Nec_F\left(s\left(F\right)\right) = \begin{cases} h(F) & \text{when } s\left(F\right) \neq X \\ 1 & \text{when } s\left(F\right) = X \end{cases}, \qquad (2.125)$$

$$\begin{aligned} Nec_F\left(\bar{s}\left(F\right)\right) &= 1 - Pos_F\left(s\left(F\right)\right) & (2.126) \\ &= 0, \end{aligned}$$

$$Pos_F\left(\emptyset\right) = Nec_F\left(\emptyset\right) = 0, \qquad (2.127)$$

$$Pos_F\left(X\right) = Nec_F\left(X\right) = 1. \qquad (2.128)$$

The interpretation is also sound under the situation of totally conflicting evidence, when $F = \emptyset$. In this case, we obtain

$$r_\emptyset\left(x\right) = 1 \text{ for all } x \in X, \qquad (2.129)$$

and, for all $A \in \mathcal{P}(X)$,

$$Pos_\emptyset\left(A\right) = \begin{cases} 1 & \text{when } A \neq \emptyset \\ 0 & \text{when } A = \emptyset \end{cases} \qquad (2.130)$$

and

$$Nec_\emptyset\left(A\right) = \begin{cases} 0 & \text{when } A \neq X \\ 1 & \text{when } A = X \end{cases}. \qquad (2.131)$$

For any fuzzy sets $F, A \in \tilde{\mathcal{P}}(X)$, we have the general formulas

$$Pos_F\left(A\right) = \sup_{x \in X} \min\left[r_F\left(x\right), A\left(x\right)\right] \qquad (2.132)$$

and

$$\begin{aligned} Nec_F\left(A\right) &= 1 - Pos_F\left(\bar{A}\right) & (2.133) \\ &= 1 - \sup_{x \in X} \min\left[r_F\left(x\right), \bar{A}\left(x\right)\right]. & (2.134) \end{aligned}$$

When $r(x) \in \{0,1\}$ for each $x \in X$, the associated possibility measure and necessity measure are called crisp. They represent, in this case, a crisp set whose characteristic function is equal to r.

When r is a *joint possibility distribution function* defined on a finite Cartesian product $X \times Y$, *marginal possibility distribution functions* r_X and r_Y are determined by the formulas

$$r_X(x) = \max_{y \in Y} r(x, y) \qquad (2.135)$$

for each $x \in X$ and

$$r_Y(y) = \max_{x \in X} r(x, y) \qquad (2.136)$$

for each $y \in Y$. The marginal bodies of evidence are viewed as noninteractive in possibility theory when

$$r(x, y) = \min[r_X(x), r_Y(y)] \qquad (2.137)$$

FIGURE 2.3 Inclusion relationships among relevant types of fuzzy measures.

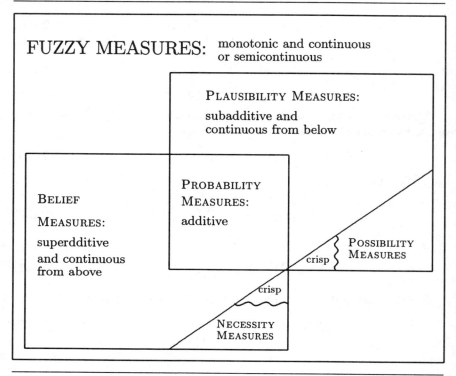

for all $x \in X$ and all $y \in Y$. Observe the analogy between these definitions and the definitions of marginal fuzzy sets, cylindric extension, and cylindric closure introduced in Sec. 2.2. Using the connection between possibility distributions and fuzzy sets, an alternative definition of noninteraction in possibility theory is the equation $r_Z(x,y) = r_{\hat{Z}}(x,y)$ for all $x \in X$ and all $y \in Y$ with $Z = X \times Y$, where r_Z and $r_{\hat{Z}}$ are possibility distributions defined in terms of a fuzzy relation on Z, and $\hat{Z}$ is the cylindric closure.

The most comprehensive coverage of possibility theory was prepared by Dubois and Prade [1988]. Other useful references are [Klir and Folger, 1988; Shafer 1985; De Cooman, 1997; De Cooman et al., 1995].

2.7 Overview of Uncertainty Theories

The relationship among the various uncertainty theories introduced in this section is summarized by a Venn diagram in Fig. 2.3. The diagram shows the inclusion of the set of measures of one type in the set of measures of another, more general type.

The most general measures are fuzzy measures (continuous or semicontinuous), which subsume all the other types. Belief measures and plausibility measures, which form the basis of evidence theory, subsume, in turn, necessity measures and possibility measures, respectively, and the latter subsume their crisp versions. Probability measures cover exactly the overlap of plausibility measures and belief measures. They overlap with neither possibility measures nor necessity measures except for one very special measure, shared by all these three types, which is defined by $m(\{x\}) = 1$ for one particular $x \in X$. This, clearly, is the only measure that represents full certainty.

Although fuzzy set theory is not explicitly shown in Fig. 2.3, it is connected with possibility theory, as explained in Sec. 2.6. Similarly, classical set theory is connected with crisp possibility theory.

3
UNCERTAINTY MEASURES

When the seemingly unique connection between uncertainty and probability theory was broken, and uncertainty began to be conceived in terms of the much broader frameworks of fuzzy set theory and fuzzy measure theory, it soon became clear that uncertainty can manifest itself in different forms. These forms represent distinct types of uncertainty. In probability theory, uncertainty is manifested only in one form.

Three types of uncertainty are now recognized in the following five theories, which are currently the only theories in which measurement of uncertainty is well established: classical set theory, fuzzy set theory, probability theory, possibility theory, and evidence theory. The three uncertainty types are: *fuzziness* (or vagueness), which results from the imprecise boundaries of fuzzy sets; *nonspecificity* (or imprecision), which is connected with sizes (cardinalities) of relevant sets of alternatives; and *strife* (or discord), which expresses conflicts among the various sets of alternatives.

It is conceivable that other types of uncertainty will be discovered when the investigation of uncertainty extends beyond the boundaries of the above mentioned five theories of uncertainty. Rather than speculating about this issue, this section is restricted to the three currently recognized types of uncertainty (and the associated information). It is shown for each of the five theories of uncertainty which uncertainty type is manifested in it and how the amount of uncertainty of that type can adequately be measured.

3.1 Nonspecificity

Hartley Function

Measurement of uncertainty (and associated information) was first conceived in terms of finite crisp sets by Hartley [1928]. Crisp sets are perhaps the simplest means by which we can express uncertainty. This kind of uncertainty emerges whenever we know that some alternative of our interest belongs to a particular set of alternatives, but we do not know which one in the set it is. To identify the alternative in question, we need information by which the uncertainty is completely removed. The amount of uncertainty associated with a set of alternatives may thus be measured by the amount of information needed to remove the uncertainty. This is the route taken by Hartley. He derived his measure of uncertainty in the following way.

Consider a finite set A of symbols that convey certain meanings in a certain context. Sequences can be formed from elements of A by successive selection. If s selections are performed then there are $|A|^s$ different potential sequences (where $|A|$ denotes the cardinality of A). The amount of information, $H(|A|^s)$, to remove the uncertainty associated with s selections should be proportional to s, that is

$$H(|A|^s) = K(|A|) \cdot s, \tag{3.1}$$

where $K(|A|)$ is a constant for the given set A that depends on $|A|$.

Consider now two sets A and B with $|A| \neq |B|$. If we perform s_1 selections from A and s_2 selections from B and yield the same number of resultant sequences in both cases then the amount of information should be the same in either case. Formally, when

$$|A|^{s_1} = |B|^{s_2}, \tag{3.2}$$

then

$$K(|A|) \cdot s_1 = K(|B|) \cdot s_2 . \tag{3.3}$$

From Eqs. (3.2) and (3.3), we obtain

$$\frac{s_2}{s_1} = \frac{\log_b |A|}{\log_b |B|} \tag{3.4}$$

and

$$\frac{s_2}{s_1} = \frac{K(|A|)}{K(|B|)} \tag{3.5}$$

respectively. Hence

$$\frac{K(|A|)}{K(|B|)} = \frac{\log_b |A|}{\log_b |B|} . \tag{3.6}$$

This equation can only be satisfied if

$$K(|A|) = c \cdot \log_b |A| , \tag{3.7}$$

where b, c are positive constants ($b > 1$, $c > 0$). Each choice of values of the constants b and c determines the unit in which uncertainty is measured. When $b = 2$ and $c = 1$, which is the most common choice, uncertainty is measured in *bits* and we obtain

$$H(A) = \log_2 |A|. \tag{3.8}$$

One bit of uncertainty is equivalent to the total uncertainty regarding the truth or falsity of one atomic proposition.

Let the set function H defined by Eq. (3.8) be called a *Hartley function*. When the Hartley function H is applied to subsets of a given universal set X, it has the form

$$H : \mathcal{P}(X) \to \mathbb{R}^+, \tag{3.9}$$

where $\mathbb{R}^+$ denotes the set of nonnegative real numbers. In this case, its range is

$$0 \leq H(A) \leq \log_2 |X|. \tag{3.10}$$

The uniqueness of the Hartley function as a measure of uncertainty (in bits) associated with finite sets of alternatives was also proven on axiomatic grounds by Rényi [1970]. Since, according to our intuition, the uncertainty depends only on the number of elements in each given set, the measure of uncertainty, H, may be viewed as a function of the form

$$H : \mathbb{N} \to \mathbb{R}^+ . \tag{3.11}$$

Using this form, Rényi characterized the measure of uncertainty with the following axioms:

(H1) Additivity — $H(n \cdot m) = H(n) + H(m)$.

(H2) Monotonicity — $H(n) \leq H(n + 1)$.

(H3) Normalization — $H(2) = 1$.

Axiom (H1) involves a set with $m \cdot n$ elements, which can be partitioned into n subsets each with m elements. A characterization of an element from the full set requires the amount $H(m \cdot n)$ of information. However, we can also proceed in two steps to characterize the element by taking advantage of the partition of the set. First, we characterize the subset to which the element belongs: the required information is $H(n)$. Then, we characterize the element within the subset: the information required is $H(m)$. These two amounts of information completely characterize an element of the full set and, hence, their sum should equal $H(m \cdot n)$. This is exactly what the axiom requires.

Axiom (H2) represents an essential and rather obvious requirement: the larger the number of alternatives, the more information is needed to characterize one of them. Axiom (H3) is needed to define the unit of information. In our case, the defined unit is the bit.

Using these axioms, Rényi established that the Hartley function is the only function that satisfies them. This is stated in the following uniqueness theorem.

Theorem 3.1 *The function $H(n) = \log_2 n$ is the only function that satisfies Axioms (H1)–(H3).*

Proof. Let n be an integer greater than 2. For each integer i define the integer $q(i)$ such that

$$2^{q(i)} \le n^i < 2^{q(i)+1}. \tag{3.12}$$

These inequalities can be written as

$$q(i) \log_2 2 \le i \log_2 n < (q(i) + 1) \log_2 2. \tag{3.13}$$

When we divide these inequalities by i and replace $\log_2 2$ with 1 we get

$$\frac{q(i)}{i} \le \log_2 n < \frac{q(i) + 1}{i} \tag{3.14}$$

and, consequently

$$\lim_{i \to \infty} \frac{q(i)}{i} = \log_2 n. \tag{3.15}$$

Let H denote a function that satisfies Axioms (H1)-(H3). Then, by Axiom (H2),

$$H(a) \le H(b) \tag{3.16}$$

for $a < b$. Combining (3.16) and (3.12) we obtain

$$H(2^{q(i)}) \le H(n^i) \le H(2^{q(i)+1}). \tag{3.17}$$

By Axiom (H1), we obtain

$$H(a^k) = k \cdot H(a). \tag{3.18}$$

If we apply this to all three terms of Eq. (3.17), we get

$$q(i) \cdot H(2) \le i \cdot H(n) \le (q(i) + 1) \cdot H(2). \tag{3.19}$$

By Axiom (H3), $H(2) = 1$ so these inequalities become

$$q(i) \le i \cdot H(n) \le q(i) + 1. \tag{3.20}$$

Dividing through by i yields

$$\frac{q(i)}{i} \le H(n) \le \frac{q(i) + 1}{i} \tag{3.21}$$

and, consequently

$$\lim_{i \to \infty} \frac{q(i)}{i} = H(n). \qquad (3.22)$$

Comparing (3.15) with (3.22), we conclude that $H(n) = \log_2 n$ for $n > 2$. Since $\log_2 2 = 1$ and $\log_2 1 = 0$ function $\log_2 n$ clearly satisfies all the axioms for $n = 1, 2$ as well.   ∎

The meaning of uncertainty measured by the Hartley function depends on the meaning of the set A. For example, when A is a set of predicted states of a variable (from the set X of all states defined for the variable), $H(A)$ is a measure of *predictive uncertainty*; when A is a set of possible diseases of a patient determined from relevant medical evidence, $H(A)$ is a measure of *diagnostic uncertainty*; when A is a set of possible answers to an unsettled historical question, $H(A)$ is a measure of *retrodictive uncertainty*; when A is a set of possible policies, $H(A)$ is a measure of *prescriptive uncertainty*. These various meanings of uncertainty measured by the Hartley function are closely connected, in one way or another, with decision-making. This connection, which applies equally well to all other types of uncertainty introduced in this chapter, has been thoroughly studied by Whittemore and Yovits [1973, 1974] and Yovits *et al.* [1981].

Observe that uncertainty expressed in terms of sets of alternatives results from the nonspecificity inherent in each set. Large sets result in less specific predictions, retrodictions, etc., than their smaller counterparts. Full specificity is obtained when only one alternative is possible. Hence, uncertainty expressed by sets of possible alternatives and measured by the Hartley function is well characterized by the term *nonspecificity*.

Consider now a situation characterized by a set A of possible alternatives (predictive, prescriptive, etc.). Assume that this set is reduced to its subset B by some action. Then, the amount of uncertainty-based information, $I_H(A, B)$, produced by the action, which is relevant to the situation, is equal to the amount of reduced uncertainty given by the difference $H(A) - H(B)$. That is,

$$I_H(A, B) = H(A) - H(B) = \log_2 \frac{|A|}{|B|} . \qquad (3.23)$$

When the action eliminates all alternatives except one (i.e., when $|B| = 1$), we obtain $I_H(A, B) = log_2|A| = H(A)$. This means that $H(A)$ may also be viewed as the amount of information needed to characterize one element of set A.

Consider now two universal sets, X and Y, and assume that a relation $R \subset X \times Y$ describes a set of possible alternatives in some situation of interest. Assume further that the domain and range of R are sets $R_X \subset X$ and $R_Y \subset Y$, respectively. Then three distinct Hartley functions are applicable, defined on the power sets of X, Y, and $X \times Y$. To identify clearly which universal set is involved in each case, it is useful (and a common practice) to write $H(X)$, $H(Y)$, $H(X,Y)$ instead of $H(R_X)$, $H(R_Y)$, $H(R)$, respectively. Functions

$$H(X) \ = \ \log_2 |R_X| \tag{3.24a}$$
$$H(Y) \ = \ \log_2 |R_Y| \tag{3.24b}$$

are called *simple uncertainties*, while function

$$H(X,Y) = \log_2 |R| \tag{3.25}$$

is called a *joint uncertainty*.

Two additional Hartley functions are defined,

$$H(X|Y) \ = \ \log_2 \frac{|R|}{|R_Y|} \tag{3.26a}$$

$$H(Y|X) \ = \ \log_2 \frac{|R|}{|R_X|} \tag{3.26b}$$

which are called *conditional uncertainties*.

Observe that the ratio $|R|/|R_Y|$ in $H(X|Y)$ represents the average number of elements of X that are possible alternatives under the condition that an element of Y has already been selected. This means that $H(X|Y)$ measures the average nonspecificity regarding alternative choices from X for all particular choices from Y. Function $H(Y|X)$ has clearly a similar meaning with the roles of sets X and Y exchanged. Observe also that the conditional uncertainties can be expressed in terms of the joint uncertainty and the two simple uncertainties:

$$H(X|Y) \ = \ H(X,Y) - H(Y), \tag{3.26c}$$
$$H(Y|X) \ = \ H(X,Y) - H(X). \tag{3.26d}$$

Furthermore,

$$H(X) - H(Y) = H(X|Y) - H(Y|X) \tag{3.27}$$

which follows immediately from Eqs. (3.26c) and (3.26d).

If possible alternatives from X do not depend on selections from Y, and vice versa, then $R = X \times Y$ and the sets X and Y are called *noninteractive*. Then, clearly,

$$H(X|Y) \ = \ H(X), \tag{3.28a}$$
$$H(Y|X) \ = \ H(Y), \tag{3.28b}$$
$$H(X,Y) \ = \ H(X) + H(Y). \tag{3.28c}$$

In all other cases, when sets X and Y are *interactive*, these equations become the inequalities

$$H(X|Y) \ < \ H(X), \tag{3.28d}$$
$$H(Y|X) \ < \ H(Y), \tag{3.28e}$$
$$H(X,Y) \ < \ H(X) + H(Y). \tag{3.28f}$$

The following symmetric function, which is usually referred to as *information transmission*, is a useful indicator of the strength of constraint between sets X and Y:

$$T_H(X,Y) = H(X) + H(Y) - H(X,Y). \tag{3.29}$$

When the sets are noninteractive, $T_H(X,Y) = 0$; otherwise, $T_H(X,Y) > 0$. Using Eqs. (3.26c) and (3.26d), $T_H(X,Y)$ can also be expressed in terms of the conditional uncertainties:

$$
\begin{aligned}
T_H(X,Y) &= H(X) - H(X|Y), &\tag{3.30a}\\
T_H(X,Y) &= H(Y) - H(Y|X). &\tag{3.30b}
\end{aligned}
$$

U-uncertainty

A natural generalization of the Hartley function from classical set theory to fuzzy set theory and possibility theory was proposed by Higashi and Klir [1983a] under the name U-*uncertainty*. For any normal fuzzy set F defined on a finite universal set X, the U-uncertainty has the form

$$U(F) = \int_0^1 \log_2 |{}^\alpha F| \; d\alpha \tag{3.31}$$

where $|{}^\alpha F|$ denotes the cardinality of the α-cut of F. Observe that $U(F)$ is a weighted average of values of the Hartley function for all the α-cuts. Each weight is a difference between the values of α of a given α-cut and the immediately preceding α-cut. When F is not normal, Higashi and Klir suggested the use of a more general form

$$U(F) = \frac{1}{h(F)} \int_0^{h(F)} \log_2 |{}^\alpha F| \; d\alpha \tag{3.32}$$

The U-uncertainty was investigated more thoroughly within possibility theory, utilizing ordered possibility distributions. In this domain, function U has the form

$$U : \mathcal{R} \to \mathbb{R}^+, \tag{3.33}$$

where $\mathcal{R}$ denotes the set of all finite and ordered possibility distributions. Given a possibility distribution

$$\mathbf{r} = \langle r_1, r_2, ..., r_n \rangle \text{ such that } 1 = r_1 \geq r_2 \geq ... \geq r_n, \tag{3.34}$$

the U-uncertainty of $\mathbf{r}$, $U(\mathbf{r})$, can be expressed by a convenient form

$$U(\mathbf{r}) = \sum_{i=2}^n (r_i - r_{i+1}) \log_2 i = \sum_{i=2}^n r_i \log_2 \left(\frac{i}{i-1} \right) \tag{3.35}$$

where $r_{n+1} = 0$ by convention [Klir and Folger, 1988]. Assume that the possibility distribution $\mathbf{r}$ in Eq. (3.34) represents the normal fuzzy set F in Eq. (3.31) in terms of Eq. (2.113). Then, it is easy to see that $U(\mathbf{r}) = U(F)$: whenever $r_i - r_{i+1} > 0$, i represents the cardinality of the α-cut with $\alpha = r_i$.

When terms on the right-hand side of Eq. (3.35) are replaced according to Eq. (2.99d), the U-uncertainty can be expressed by another formula:

$$U(\mathbf{m}) = \sum_{i=2}^{n} m_i \log_2 i \qquad (3.36)$$

where $\mathbf{m} = \langle m_1, m_2, ..., m_n \rangle$ represents the basic probability assignment corresponding to $\mathbf{r}$.

Given the possibility distribution function $\mathbf{r}_F$ associated with a particular fuzzy set F (not necessarily normal) via Eq. (2.121), the basic probability assignment $\mathbf{m}_F$ used for calculating the weighted average of the Hartley function for focal elements is given by Eq. (2.99d). Since focal elements are in this case identified with the α–cuts of F, the U–uncertainty $U(\mathbf{r}_F)$, which expresses this weighted average is given by the formula

$$U(\mathbf{r}_F) = \int_0^{h(F)} \log_2 |{}^\alpha F| \, d\alpha + (1 - h(F)) \log_2 |X| \,. \qquad (3.37)$$

This formula may also be adopted for calculating the nonspecificity of F. That is,

$$U(F) = \int_0^{h(F)} \log_2 |{}^\alpha F| \, d\alpha + (1 - h(F)) \log_2 |X| \,. \qquad (3.38)$$

Observe that this formula compares to (3.32) when F is normal, but it differs from (3.32) when F is subnormal. While formula (3.32) involves the scaling of F, for which there is little justification, formula (3.38) does not involve any modification of F. Moreover, formula (3.38) is based on a well–justified connection (2.132) between subnormal fuzzy sets and possibility distributions, as discussed in Sec. 2.6. Hence, function U defined by (3.38) rather than (3.32) should be adopted to measure the uncertainty of fuzzy sets.

When a possibility distribution $\mathbf{r}$ is derived from a subnormal fuzzy set, then $r_1 < 1$ and $\sum_{i=1}^{n} m_i = r_1$. In this case, formulas (3.35) and (3.36) must be replaced with more general formulas

$$\begin{aligned} U(\mathbf{r}) &= \sum_{i=2}^{n} (r_i - r_{i+1}) \log_2 i + (1 - r_1) \log_2 n \qquad (3.39) \\ &= \sum_{i=2}^{n} r_i \log \frac{i}{i-1} + (1 - r_1) \log_2 n \end{aligned}$$

and

$$U(\mathbf{m}) = \sum_{i=2}^{n} m_i \log i + (1 - r_1) \log_2 n, \qquad (3.40)$$

respectively.

Consider now two universal sets, X and Y, and a joint possibility distribution $\mathbf{r}$ defined on $X \times Y$. Adopting the notation introduced for the Hartley function, let $U(X,Y)$ denote the joint U-uncertainty, and let $U(X)$ and $U(Y)$ denote simple U-uncertainties defined on the marginal possibility distributions $\mathbf{r}_X$ and $\mathbf{r}_Y$, respectively. Then, we have

$$U(X) = \sum_{A \in \mathcal{F}_X} m_X(A) \log_2 |A|, \qquad (3.41)$$

$$U(Y) = \sum_{B \in \mathcal{F}_Y} m_Y(B) \log_2 |B|, \qquad (3.42)$$

$$U(X,Y) = \sum_{A \times B \in \mathcal{F}} m(A \times B) \log_2 |A \times B|, \qquad (3.43)$$

where $\mathcal{F}_X$, $\mathcal{F}_Y$, and $\mathcal{F}$ are sets of focal elements induced by m_X, m_Y, and m, respectively. Furthermore, we define conditional U-uncertainties, $U(X|Y)$ and $U(Y|X)$, as the following generalizations of the corresponding conditional Hartley functions:

$$U(X|Y) = \sum_{A \times B \in \mathcal{F}} m(A \times B) \log_2 \frac{|A \times B|}{|B|}, \qquad (3.44)$$

$$U(Y|X) = \sum_{A \times B \in \mathcal{F}} m(A \times B) \log_2 \frac{|A \times B|}{|A|}, \qquad (3.45)$$

Observe that the term $|A \times B|/|B|$ in Eq. (3.44) represents for each focal element $A \times B$ in $\mathcal{F}$ the average number of elements of A that remain possible alternatives under the condition that an element of Y has already been selected. Expressing $U(X|Y)$ in the form of Eq. (3.31), we have

$$
\begin{aligned}
U(X|Y) &= \int_0^1 \log_2 \frac{|{}^{\alpha}(A \times B)|}{|{}^{\alpha}B|} \, d\alpha \qquad (3.46) \\
&= \int_0^1 \log_2 |{}^{\alpha}(A \times B)| \, d\alpha - \int_0^1 \log_2 |{}^{\alpha}B| \, d\alpha \\
&= U(X,Y) - U(Y)
\end{aligned}
$$

where $A \times B$ is a fuzzy set with membership function $(A \times B)(x,y) = \min[A(x), B(y)]$. This equation is clearly a generalization of Eq. (3.26c) from crisp sets to fuzzy sets. In a similar way, a generalization of Eq. (3.26d) can be derived. Using these two generalized equations, it can easily be shown that also Eqs. (3.27) - (3.28f) are valid for the U-uncertainty

[Higashi and Klir, 1983a; Klir and Folger, 1988]. Furthermore, information transmission can be defined for the U-uncertainty by Eq. (3.29) and, then, Eqs. (3.30a) and (3.30b) are also valid in the generalized framework.

Any proof of uniqueness of the U-uncertainty as a measure of uncertainty and information for fuzzy sets or possibility theory uses a subset of the following group of axioms (stated in terms of possibility distributions):

(U1) **Additivity** — $U(\mathbf{r}) = U(\mathbf{r}_X) + U(\mathbf{r}_Y)$ when $\mathbf{r}$ is a noninteractive possibility distribution on $X \times Y$.

(U2) **Subadditivity** — It is expressed by the inequality $U(\mathbf{r}) \leq U(\mathbf{r}_X) + U(\mathbf{r}_Y)$, where $\mathbf{r}$ is any possibility distribution on $X \times Y$.

(U3) **Expansibility** — When components of zero are added to a given possibility distribution, the value of the U-uncertainty does not change.

(U4) **Symmetry** — $U(\mathbf{r}) = U(\mathbf{q})$, where $\mathbf{q}$ is any permutation of $\mathbf{r}$.

(U5) **Continuity** — U is a continuous function.

(U6) **Monotonicity** — For any pair $^1\mathbf{r}$, $^2\mathbf{r}$ of possibility distributions of the same length, if $^1\mathbf{r} \leq {}^2\mathbf{r}$ then $U(^1\mathbf{r}) \leq U(^2\mathbf{r})$.

(U7) **Measurement unit** — To use bits as measurement units, it is required that $U(1,1) = 1$.

(U8) **Range** — Within a given universal set X, $U(r) \in [0, log_2|X|]$, where the minimum and maximum are obtained for $\mathbf{r} = \langle 1, 0, ..., 0 \rangle$ and $\mathbf{r} = \langle 1, 1, ..., 1 \rangle$, respectively.

(U9) **Branching** — For every possibility distribution $\mathbf{r} = \langle r_1, r_2, ..., r_n \rangle$ of any length n,

$$U(r_1, r_2, ..., r_n) = U(r_1, r_2, ..., r_{k-2}, r_k, r_k, r_{k+1}, ..., r_n) \quad (3.47)$$

$$\left(+(r_{k-2} - r_k)U\left(1_{k-2}, \frac{r_{k-1} - r_k}{r_{k-2} - r_k}, 0_{n-k+1}\right) \right.$$
$$\left. -(r_{k-2} - r_k)U(1_{k-2}, 0_{n-k+2}) \right)$$

compensate for what was skipped   *crisp*

for each $k = 3, 4, ..., n$, where, for any given integer i, $\mathbf{1}_i$ denotes a sequence of i ones and $\mathbf{0}_i$ denotes a sequence of i zeroes.

The branching axiom, which can also be formulated in other forms [Klir and Mariano, 1987; Klir and Folger, 1988], needs some explanation. This axiom basically requires that the U-uncertainty has the capability of measuring possibilistic nonspecificity in two ways. It can be measured either directly for the given possibility distribution or indirectly by adding U-uncertainties associated with a combination of possibility distributions that reflect a two-stage measuring process. In the first stage of measurement, the distinction between two neighboring components, r_{k-1} and r_k, is ignored (r_{k-1} is replaced with r_k) and the U-uncertainty of the resulting, less refined possibility distribution is calculated. In the second stage, the U-uncertainty is calculated in a local frame of reference, which is defined by a possibility distribution that distinguishes only between the two neighboring possibility values that are not distinguished in the first stage of measurement. The U-uncertainty calculated in the local frame must be scaled back to the original frame by a suitable weighting factor. The sum of the two U-uncertainties obtained by the two stages of measurement is equal to the total U-uncertainty of the given possibility distribution.

The first term on the right-hand side of Eq. (3.47) represents the U-uncertainty obtained in the first stage of measurement. The remaining two terms represent the second stage, associated with the local U-uncertainty. The first of these two terms expresses the loss of uncertainty caused by ignoring the component r_{k-1} in the given possibility distribution, but it introduces some additional U-uncertainty equivalent to the uncertainty of a crisp possibility distribution with $k-2$ components. This additional U-uncertainty is excluded by the last term in Eq. (3.47). That is, the local uncertainty is expressed by the last two terms in Eq. (3.47).

The meaning of the branching property of the U-uncertainty is illustrated by Fig. 3.1. Four hypothetical possibility distributions involved in Eq. (3.47) are shown in the figure and in each of them the local frame involved in the branching is indicated.

It was proven by Klir and Mariano [1987] that the U-uncertainty is a unique measure of possibilistic nonspecificity when expansibility, monotonicity, additivity, branching, and appropriate normalization (designating bits as measurement units) are chosen as axioms.

Theorem 3.2 *The U-uncertainty is the only function that satisfies the axioms of expansibility, monotonicity, additivity, branching, and measurement unit.*

The proof is presented in the Appendix.

Although other axioms could possibly be used for proving the uniqueness of function U, no such attempt has, thus far, been made.

The question of how to measure uncertainty and information for continuous possibility distributions was investigated by Ramer [1990b]. He showed that the U-uncertainty is not directly applicable to continuous possibility distributions, but it can be employed in a relative form

$$W(\mathbf{r}, \mathbf{s}) = \sum_{i=1}^{n} (s_i - r_i) \log_2 \frac{i}{i-1} \qquad (3.48)$$

which involves two possibility distributions $\mathbf{r} = \langle r_1, r_2, ..., r_n \rangle$ and $\mathbf{s} = \langle s_1, s_2, ..., s_n \rangle$ such that $r_i \leq s_i$ for all $i = 1, 2, ..., n$. When W is restricted

FIGURE 3.1 Possibility distributions involved in the branching property.

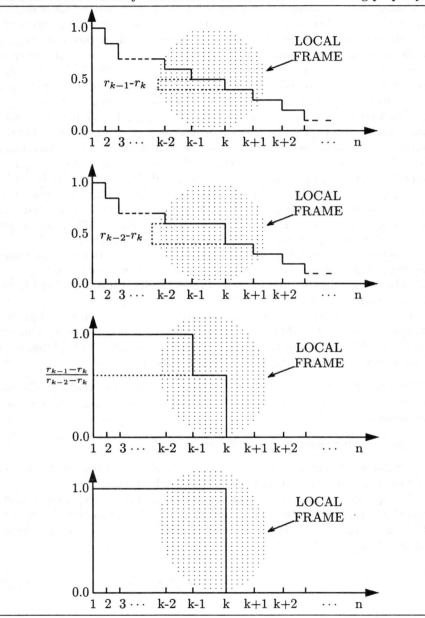

to $\mathbf{s} = \langle 1, 1, ..., 1 \rangle = \mathbf{1}_n$, we obtain

$$W(\mathbf{r}) = \log_2 n - U(r), \qquad (3.49)$$

where $W(\mathbf{r})$ is a shorthand symbol for $W(\mathbf{r}, \mathbf{1}_n)$. This function has similar properties as the U-uncertainty, except that all inequality signs are reversed. Ramer showed that

$$W(\mathbf{r}) = \int_0^1 \frac{1 - \tilde{r}(x)}{x} dx \qquad (3.50)$$

is the continuous counterpart of Eq. (3.49), where $\tilde{r}$ denotes the decreasing rearrangement [Hardy et al., 1934] of the possibility distribution function $\mathbf{r}$.

Nonspecificity in Evidence Theory

Once the U-uncertainty was well established as a possibilistic generalization of the Hartley function, Dubois and Prade [1985a] showed that it can be further generalized to measure nonspecificity in evidence theory. To distinguish the U-uncertainty from its more general counterpart in evidence theory, let the latter function be denoted by N. It is defined by the formula

$$N(m) = \sum_{A \in \mathcal{F}} m(A) \log_2 |A| \qquad (3.51)$$

where $\langle \mathcal{F}, m \rangle$ is an arbitrary body of evidence.

It was proven by Ramer [1987] that function N (or N-uncertainty) defined by Eq. (3.51) is the only function that satisfies additivity for noninteractive marginal bodies of evidence, subadditivity for interactive ones, normalization based on choosing bits as measurement units ($N(m) = 1$ when $m(A) = 1$ and $|A| = 2$), and the following two properties:

(N1) Symmetry — N is invariant with respect to permutations of values of the basic probability assignment within each group of subsets of X that have equal cardinalities.

(N2) Branching — $N(m) = N(m_1) + N(m_2)$ for any three bodies of evidence $\langle \mathcal{F}, m \rangle$, $\langle \mathcal{F}_1, m_1 \rangle$, and $\langle \mathcal{F}_2, m_2 \rangle$ such that

$$\begin{aligned} \mathcal{F} &= \{A, B, C, ...\}, \qquad (3.52) \\ \mathcal{F}_1 &= \{A_1, B, C, ...\}, \\ \mathcal{F}_2 &= \{A, B_1, C_1, ...\}, \end{aligned}$$

where $A_1 \subset A, B_1 \subset B, C_1 \subset C, ..., |A_1| = |B_1| = |C_1| = \cdots = 1$, and

$$\begin{aligned} m(A) &= m_1(A_1) = m_2(A), \qquad (3.53) \\ m(B) &= m_1(B) = m_2(B_1), \\ m(C) &= m_1(C) = m_2(C_1), \text{ etc.} \end{aligned}$$

Function N is clearly a weighted average of the Hartley function for all focal elements. The weights are values of the basic probability assignment. For each focal element A, $m(A)$ indicates the degree of evidence focusing on A, while $log_2|A|$ indicates the lack of specificity of this evidential claim. The larger the value of $m(A)$, the stronger the evidence; the larger the set A (and $log_2|A|$), the less specific the evidence. Consequently, it is reasonable to view function N as a *measure of nonspecificity*.

The range of function N is, as expected, $[0, log_2|X|]$. The minimum, $N(m) = 0$, is obtained when $m(\{x\}) = 1$ for some $x \in X$ (no uncertainty); the maximum, $N(m) = \log_2 |X|$, is reached for $m(X) = 1$ (total ignorance). It can easily be shown that Eqs. (3.26c) - (3.30b) remain valid when the Hartley function is generalized to the N-uncertainty.

When focal elements are nested and we employ the notation introduced in connection with Eqs. (2.99c) and (2.99d), $N(m)$ can be written as

$$N(m) = \sum_{i=1}^{n} m(A_i) \log_2 |A_i|, \qquad (3.54)$$

Using Eq. (2.99d), this form is directly convertible into an equivalent form of the U-uncertainty,

$$U(r) = \sum_{i=1}^{n} (r(x_i) - r(x_{i+1})) \log_2 |A_i|, \qquad (3.55)$$

where $x_i \in A_i$, $x_{i+1} \in A_{i+1}$, and $r(x_{n+1}) = 0$.

This demonstrates that the U-uncertainty is a special case (restricted to possibility theory) of the general measure of nonspecificity N.

Since focal elements in probability measures are singletons, $|A| = 1$ and $\log_2 |A| = 0$ for each focal element. Consequently, $N(m) = 0$ for every probability measure. That is, probability theory is not capable of incorporating nonspecificity, one of the basic types of uncertainty.

In addition to the mentioned properties of the N-uncertainty, which make it well justified as a measure of nonspecificity in evidence theory, the function has some other, surprisingly strong properties. As shown by Klir and Wierman [1987], for example, the N-uncertainty of a joint body of evidence obtained from nested marginal bodies of evidence by the rules of possibility theory (applying the minimum operator to marginal possibility distributions) is exactly the same as the N-uncertainty of a totally different joint body of evidence obtained from the same marginals by the rules of evidence theory. The N-uncertainty is also additive not only for noninteractive bodies of evidence, but also for any joint body of evidence on $X \times Y$ whose only focal elements are Cartesian products of some subsets of X and some subsets of Y.

Assume that sets X, Y, and Z, with $Z = X \times Y$ are all finite sets and that $\hat{Z}$ represents the cylindric closure.

Lemma 3.3 *The Cartesian product of the α–cuts of two fuzzy sets defined on X and Y is equal to the α–cut of their cylindric closure.*

Proof. First we will show that the α-cut of the cylindric closure contains the Cartesian product of the α-cuts of the marginals. Fix α, if $A_X(x) \geq \alpha$ and $A_Y(y) \geq \alpha$ then

$$A_{\hat{Z}}(x, y) = \min\left[A_X(x), A_Y(y)\right] \geq \alpha. \tag{3.56}$$

This implies that if $x \in {}^{\alpha}A_X$ and $y \in {}^{\alpha}A_Y$ then $\langle x, y \rangle \in {}^{\alpha}A_{\hat{Z}}$. Thus

$$^{\alpha}A_X \times {}^{\alpha}A_Y \subseteq {}^{\alpha}A_{\hat{Z}}. \tag{3.57}$$

The reverse inclusion comes from noting that if $z \in {}^{\alpha}A_{\hat{Z}}$ then $z = \langle x, y \rangle$ for some $x \in x$ and $y \in Y$. But $z \in {}^{\alpha}A_{\hat{Z}}$ also implies that $A_{\hat{Z}}(z) \geq \alpha$, which means $\min[A_X(x), A_Y(y)] \geq \alpha$ and, of course, if the minimum of two values is greater than or equal to α, then both values individually are greater than or equal to α. So both $x \in {}^{\alpha}A_X$ and $y \in {}^{\alpha}A_Y$. Thus

$$^{\alpha}A_X \times {}^{\alpha}A_Y \supseteq {}^{\alpha}A_{\hat{Z}}. \tag{3.58}$$

∎

Theorem 3.4 *The sum of the U-uncertainties $U(A_X)$ and $U(A_Y)$ is equal to the U-uncertainty of the cylindric closure of A_X and A_Y. That is,*

$$U(A_{\hat{Z}}) = U(A_X) + U(A_Y) \tag{3.59}$$

or

$$\int_0^1 \log_2 \left|{}^{\alpha}A_{\hat{Z}}\right| d\alpha = \int_0^1 \log_2 \left|{}^{\alpha}A_X\right| d\alpha + \int_0^1 \log_2 \left|{}^{\alpha}A_Y\right| d\alpha. \tag{3.60}$$

Proof.

$$^{\alpha}A_{\hat{Z}} = {}^{\alpha}A_X \times {}^{\alpha}A_Y \tag{3.61}$$

by Lemma (3.3). The sets involved are crisp so that

$$\left|{}^{\alpha}A_{\hat{Z}}\right| = \left|{}^{\alpha}A_X\right| \cdot \left|{}^{\alpha}A_Y\right| \tag{3.62}$$

and further

$$\log_2 \left|{}^{\alpha}A_{\hat{Z}}\right| = \log_2 \left|{}^{\alpha}A_X\right| + \log_2 \left|{}^{\alpha}A_Y\right|. \tag{3.63}$$

This is true for any value of α so if we integrate both sides of the equality over alpha from zero to one the proposition is proved. ∎

This result is not surprising, since the cylindric closure of the fuzzy sets A_X and A_Y is a noninteractive fuzzy set with marginals A_X and A_Y. If this were false it would violate the Axiom (U1). What is surprising is that we can use the marginal possibility distributions $\mathbf{r}_X$ and $\mathbf{r}_Y$ to create marginal basic probability assignments m_X and m_Y and then create a joint basic probability assignment m_Z and prove that

$$
\begin{aligned}
U(\mathbf{r}_Z) &= N(m_Z) \\
&= N(m_X) + N(m_Y) \\
&= U(\mathbf{r}_X) + U(\mathbf{r}_Y) \\
&= U(\mathbf{r}_{\hat{z}}) \\
&= N(m_{\hat{z}})
\end{aligned}
\tag{3.64}
$$

where $m_{\hat{z}}$ is created using a product operator in the realm of evidence theory.

Theorem 3.5 *If $\mathbf{r}_Z$ is a noninteractive joint possibility distribution then*

$$
\begin{aligned}
U(\mathbf{r}_Z) &= N(m_Z) \\
&= N(m_X) + N(m_Y) \\
&= U(\mathbf{r}_X) + U(\mathbf{r}_Y) \\
&= U(\mathbf{r}_{\hat{z}}) \\
&= N(m_{\hat{z}})
\end{aligned}
\tag{3.65}
$$

Proof. The proof of

$$
U(\mathbf{r}_Z) = U(\mathbf{r}_X) + U(\mathbf{r}_Y) = U(\mathbf{r}_{\hat{z}})
\tag{3.66}
$$

is identical to the proof of Theorem 3.4. As mentioned in Sec. 2.6 a possibility distribution induces a basic probability assignment on a sequence of nested focal sets $A_i = \{x_1, x_2, \ldots, x_i\}$ with $m_i = m(A_i) = r_i - r_{i+1}$ for all $i = 1, 2, \ldots, n$ and where $r_{n+1} = 0$ by convention. Therefore we can use the possibility distributions $\mathbf{r}_X$, $\mathbf{r}_Y$, and $\mathbf{r}_Z$ to define

$$
\begin{array}{lll}
m_{X,i} = m_X(A_i) = r_{X,i} - r_{X,i+1} & \text{with } A_i \subseteq X \\
m_{Y,j} = m_Y(B_j) = r_{Y,j} - r_{Y,j+1} & \text{with } B_j \subseteq Y \\
m_{Z,k} = m_Z(C_k) = r_{Z.k} - r_{Z,k+1} & \text{with } C_k \subseteq Z
\end{array}
\tag{3.67}
$$

Since $U(\mathbf{r}) = N(m)$ (due to $m_i = r_i - r_{i+1}$ and $|A_i| = i$, the definitions are intentionally identical) for any corresponding possibility distribution and basic probability assignment, we know that

$$
\begin{aligned}
U(\mathbf{r}_X) &= N(m_X), & \text{(3.68a)} \\
U(\mathbf{r}_Y) &= N(m_Y), & \text{(3.68b)} \\
U(\mathbf{r}_Z) &= N(m_Z), & \text{(3.68c)}
\end{aligned}
$$

and obviously

$$
U(\mathbf{r}_Z) + U(\mathbf{r}_Y) = N(m_X) + N(m_Y)
\tag{3.69}
$$

All that is needed to finish the proof is to show that

$$N(m_{\hat{Z}}) = N(m_X) + N(m_Y) \quad . \tag{3.70}$$

To make matters simpler, let us again indicate the cylindric closure with $\hat{Z}$; then,

$$m_{\hat{Z}}(A \times B) = m_X(A) \cdot m_Y(B) \tag{3.71}$$

and we have

$$N(m_{\hat{Z}}) = \sum_{C \subseteq \mathcal{F}_{\hat{Z}}} m_{\hat{Z}}(C) \log_2 |C| \tag{3.72}$$

but, by construction, $\mathcal{F}_{\hat{Z}}$ only contains sets of the form $A \times B$ with $A \subseteq \mathcal{F}_X$ and $B \subseteq \mathcal{F}_Y$. Thus

$$
\begin{aligned}
N(m_{\hat{Z}}) &= \sum_{C \subseteq \mathcal{F}_Z} m_{\hat{Z}}(C) \log_2 |C| \tag{3.73}\\
&= \sum_{A \times B \subseteq \mathcal{F}_{\hat{Z}}} m_{\hat{Z}}(A \times B) \log_2 |A \times B|\\
&= \sum_{A \subseteq \mathcal{F}_X} \sum_{B \subseteq \mathcal{F}_Y} m_{\hat{Z}}(A \times B) \log_2 |A \times B|\\
&= \sum_{A \subseteq \mathcal{F}_X} \sum_{B \subseteq \mathcal{F}_Y} m_X(A) \cdot m_Y(B) \left(\log_2 |A| + \log_2 |B|\right)\\
&= \sum_{A \subseteq \mathcal{F}_X} \sum_{B \subseteq \mathcal{F}_Y} m_X(A) \cdot m_Y(B) \log_2 |A|\\
&\quad + \sum_{A \subseteq \mathcal{F}_X} \sum_{B \subseteq \mathcal{F}_Y} m_X(A) \cdot m_Y(B) \log_2 |B|\\
&= \sum_{A \subseteq \mathcal{F}_X} m_X(A) \log_2 |A| + \sum_{B \subseteq \mathcal{F}_Y} m_Y(B) \log_2 |B|\\
&= N(m_X) + N(m_Y)
\end{aligned}
$$

since basic probability assignments sum to one. ■

Table 3.1 shows a simple example that demonstrates the additivity of the nonspecificity measure.

Nonspecificity of Sets in n-Dimensional Euclidean Space

The Hartley function H is applicable to measure uncertainty only of finite sets. Since the U–uncertainty and the N–uncertainty are based on the Hartley function, they are subjects to the same restriction. Since fuzzy sets and fuzzy relations are often defined by continuous membership functions on the n-dimensional Euclidean space, this restriction is rather unsettling.

TABLE 3.1 Additivity of nonspecificity for noninteractive evidence.
(a) Noninteractive joint possibility distribution on
$$Z = X \times Y = \{a_1, a_2\} \times \{b_1, b_2\}$$
(b) Marginal X possibility distribution
(c) Marginal X basic probabiltiy assignment from (b)
(d) Marginal Y possibility distribution
(e) Marginal Y basic probabiltiy assignment from (d)
(f) Marginal Z basic probabiltiy assignment from (a)
(g) Cylindric closure of (c) and (e).

(a)	$\mathbf{r}_Z = \mathbf{r}_{\hat{Z}}$
$z_1 = \langle a_1, b_1 \rangle$	1.0
$z_2 = \langle a_1, b_2 \rangle$	0.7
$z_3 = \langle a_2, b_1 \rangle$	0.6
$z_4 = \langle a_2, b_2 \rangle$	0.6
U	1.3

(b)	$\mathbf{r}_X$
a_1	1
a_2	0.6
U	0.6

$$\mathbf{r}_X(x) = \max_{y \in Y} \mathbf{r}_Z(x, y)$$

(c)	m_X
$\{a_1\}$	0.4
$\{a_1, a_2\}$	0.6
N	0.6

$$m_X(\{a_1, \ldots, a_i\}) = \mathbf{r}_{X,i} - \mathbf{r}_{X,i+1}$$

(d)	$\mathbf{r}_Y$
b_1	1
b_2	0.7
U	0.7

$$\mathbf{r}_Y(y) = \max_{x \in X} \mathbf{r}_Z(x, y)$$

(e)	m_Y
$\{b_1\}$	0.3
$\{b_1, b_2\}$	0.7
N	0.7

$$m_Y(\{b_1, \ldots, b_i\}) = \mathbf{r}_{Y,i} - \mathbf{r}_{Y,i+1}$$

(f)	m_Z
$\{\langle a_1, b_1 \rangle\}$	0.3
$\{\langle a_1, b_1 \rangle, \langle a_1, b_2 \rangle\}$	0.1
$\{\langle a_1, b_1 \rangle, \langle a_1, b_2 \rangle, \langle a_2, b_1 \rangle\}$	0.0
Z	0.6
N	1.3

$$m_Z(C_i) = \mathbf{r}_{Z,i} - \mathbf{r}_{Z,i+1}$$

(g)	$m_{\hat{Z}}$
$\{\langle a_1, b_1 \rangle\}$	0.12
$\{\langle a_1, b_1 \rangle, \langle a_1, b_2 \rangle\}$	0.28
$\{\langle a_1, b_1 \rangle, \langle a_2, b_1 \rangle\}$	0.18
Z	0.42
N	1.3

$$m_{\hat{Z}}(A \times B) = m_Y(A) \cdot m_Y(B)$$

The potential of measuring nonspecificity of sets in the n-dimensional Euclidean space was investigated by Klir and Yuan [1995b]. In the following, we describe the considerations and results they obtained.

Let X denote a universal set of concern that is assumed to be a bounded and convex subset of $\mathbb{R}^n$ for some finite $n \geq 1$, and let HL denote a function, called a *Hartley-like function*, of the form

$$HL : \mathcal{C} \to \mathbb{R}^+ \tag{3.74}$$

where $\mathcal{C}$ is the family of all convex subsets of X, which for $A \in \mathcal{C}$ is expected to measure the nonspecificity of A. Thus function HL must satisfy the following axiomatic requirements:

(HL1) Range — For each $A \in \mathcal{C}$, $HL(A) \in [0, \infty)$, where $HL(A) = 0$ iff $A = \{x\}$ for some $x \in X$.

(HL2) Monotonicity — For all $A, B \in \mathcal{C}$, if $A \subseteq B$, then

$$HL(A) \leq HL(B). \tag{3.75}$$

(HL3) Subadditivity — For each $A \in \mathcal{C}$,

$$HL(A) \leq \sum_{i=1}^{n} HL(A_i), \tag{3.76}$$

where A_i denotes the one–dimensional projection of A to dimension i in some coordinate system.

(HL4) Additivity — For all $A \in \mathcal{C}$, such that $A = \times_{i=1}^{n} A_i$, where A_i has the same meaning as in (HL3),

$$HL(A) = \sum_{i=1}^{n} HL(A_i). \tag{3.77}$$

(HL5) Coordinate invariance — Function HL does not change under isometric transformations of the coordinate space.

(HL6) Continuity — Function HL is a continuous function.

As a candidate for a Hartley-like function, Klir and Yuan [1995b] proposed the function defined by the formula

$$HL(A) = \min_{t \in T} \ln \left[\prod_{i=1}^{n} [1 + \mu(A_{i_t})] + \mu(A) - \prod_{i=1}^{n} [\mu(A_{i_t})] \right], \tag{3.78}$$

where μ denotes the Lebesgue measure, T denotes the set of all transformations from one orthogonal coordinate system to another, and A_{i_t} denotes the ith projection of A within the coordinate system t. They also established the following properties of this function pertaining to the axiomatic requirements (HL1)–(HL6).

It is evident that the proposed function is continuous, invariant with respect to isometric transformations of the coordinate system, and that it satisfies the required range. The monotonicity of the function follows from the corresponding monotonicity of the Lebesgue measure, and its subadditivity is demonstrated as follows: for any $A \in \mathcal{C}$,

$$
\begin{aligned}
HL(A) \;&=\; \min_{t \in T} \ln \left[\prod_{i=1}^{n} [1 + \mu(A_{i_t})] + \mu(A) - \prod_{i=1}^{n} [\mu(A_{i_t})] \right] \quad (3.79) \\
&\leq\; \min_{t \in T} \ln \left[\prod_{i=1}^{n} [1 + \mu(A_{i_t})] \right] , \\
&=\; \min_{t \in T} \sum_{i=1}^{n} \ln [1 + \mu(A_{i_t})] , \\
&=\; \sum_{i=1}^{n} HL(A_i) .
\end{aligned}
$$

It remains to show that the proposed function is additive, to be fully justified as a general measure of nonspecificity of convex subsets of $\mathbb{R}^n$ for any finite $n \geq 1$.

To prove that the proposed function is additive, we must prove that

$$
HL(A) = \sum_{i=1}^{n} HL(A_i). \qquad (3.80)
$$

for any $A \in \mathcal{C}$, such that $A = \times_{i=1}^{n} A_i$. It has already been shown that

$$
HL(A) \leq \sum_{i=1}^{n} HL(A_i), \qquad (3.81)
$$

for any $A \in \mathcal{C}$. Hence it remains to prove that

$$
HL(A) \geq \sum_{i=1}^{n} HL(A_i), \qquad (3.82)
$$

when $A = \times_{i=1}^{n} A_i$. This, in turn, amounts to proving that for any rotation of the set A,

$$
\prod_{i=1}^{n} [1 + \mu(A_i)] + \mu(A) - \prod_{i=1}^{n} [\mu(A_{i_t})] \geq \prod_{i=1}^{n} [1 + \mu(A_{i_t})] . \qquad (3.83)
$$

At this time, it is still a conjecture that this inequality holds. However, this conjecture has such a strong support, on several distinct grounds, that its validity is very likely. Klir and Yuan [1995b] present the following support.

Generally, in the n–dimensional space $\mathbb{R}^n$, any rotation can be represented by the orthogonal matrix

$$
\mathbf{I} = \begin{bmatrix}
\cos \alpha_{11} & \cos \alpha_{12} & \cdots & \cos \alpha_{1n} \\
\cos \alpha_{21} & \cos \alpha_{22} & \cdots & \cos \alpha_{2n} \\
\vdots & \vdots & \ddots & \vdots \\
\cos \alpha_{n1} & \cos \alpha_{n2} & \cdots & \cos \alpha_{nn}
\end{bmatrix} ,
\tag{3.84}
$$

where the parameters satisfy the following properties:

1. $\sum_{i=1}^{n} \cos^2 \alpha_{ij} = 1, \quad \forall j \in \mathbb{N}_n$, and
   $\sum_{j=1}^{n} \cos^2 \alpha_{ij} = 1, \quad \forall i \in \mathbb{N}_n$.

2. $\sum_{k=1}^{n} \cos \alpha_{ik} \cos \alpha_{jk} = 0, \quad \forall i, j \in \mathbb{N}_n$, and
   $\sum_{k=1}^{n} \cos \alpha_{ki} \cos \alpha_{kj} = 0, \quad \forall i, j \in \mathbb{N}_n$.

For each given rotation defined by matrix $\mathbf{I}$, an arbitrary point

$$
\mathbf{x} = \langle x_1, x_2, \ldots, x_n \rangle^T
\tag{3.85}
$$

in $\mathbb{R}^n$ is transformed to the point

$$
\mathbf{x}' = \langle x'_1, x'_2, \ldots, x'_n \rangle^T
\tag{3.86}
$$

by the matrix equation

$$
\mathbf{x}' = \mathbf{I}\,\mathbf{x}
\tag{3.87}
$$

That is,

$$
\begin{aligned}
x'_1 &= x_1 \cos \alpha_{11} + x_2 \cos \alpha_{12} + \ldots + x_n \cos \alpha_{1n} \\
x'_2 &= x_1 \cos \alpha_{21} + x_2 \cos \alpha_{22} + \ldots + x_n \cos \alpha_{2n}
\end{aligned}
\tag{3.88}
$$

$$
\vdots
$$

$$
x'_n = x_1 \cos \alpha_{n1} + x_2 \cos \alpha_{n2} + \ldots + x_n \cos \alpha_{nn}
$$

Let us consider, without any loss of generality, that

$$
A = \times_{i=1}^{n} [0, a_i]
\tag{3.89}
$$

for some $a_i \in \mathbb{R}$, $i \in \mathbb{N}_n$. Then, the ith projection of this set subjected to rotation defined by the matrix $\mathbf{I}$ is the set

$$
A_i = \{ x'_i \mid \mathbf{x}' = \mathbf{I}\,\mathbf{x}, \quad \forall \mathbf{x} \in A \}
\tag{3.90}
$$

for any $i \in \mathbb{N}_n$. The Lebesgue measure of the projection is

$$
\mu(A_i) = \max \{ |x'_i - y'_i| \mid x, y \in A \} .
\tag{3.91}
$$

That is

$$\mu(A_i) = \max\left\{ \left| \sum_{j=1}^{n} (x_j - y_j)\cos\alpha_{ij} \right| \mid x, y \in A \right\}.$$
(3.92)

for any $i \in \mathbb{N}_n$. Since this maximum must be reached by two vertices of the set A, the Lebesgue measure of the projection can be rewritten by

$$\mu(A_i) = \sum_{k=1}^{n} a_k \left|\cos\alpha_{ik}\right|$$
(3.93)

for any $i \in \mathbb{N}_n$.

Using the last formula, let us examine some aspects of the proposed function pertaining to its additivity. First let us present the following two basic properties.

(a) $\displaystyle\sum_{i=1}^{n} \mu(A_i) \geq \sum_{i=1}^{n} a_i.$

This is because

$$
\begin{aligned}
\sum_{i=1}^{n} \mu(A_i) &= \sum_{i=1}^{n}\sum_{k=1}^{n} a_k \left|\cos\alpha_{ik}\right| & (3.94) \\
&= \sum_{k=1}^{n}\sum_{i=1}^{n} a_k \left|\cos\alpha_{ik}\right| \\
&\geq \sum_{k=1}^{n}\sum_{i=1}^{n} a_k \cos^2\alpha_{ik} \\
&= \sum_{k=1}^{n} a_k \sum_{i=1}^{n} \cos^2\alpha_{ik} \\
&= \sum_{k=1}^{n} a_k
\end{aligned}
$$

(b) $\displaystyle\prod_{i=1}^{n} \mu(A_i) \geq \prod_{i=1}^{n} a_i.$

This is because of the fact that the Lesbesgue measure of the set is less than or equal to the Lesbesgue measure of the Cartesian product of its one–dimensional projections.

The case of Sets with Equal Edges

Let $a_i = a$ for all $i \in \mathbb{N}_n$. Then, we have

$$\mu(A_i) = \sum_{k=1}^{n} a_k \left|\cos\alpha_{ik}\right|$$
(3.95)

FIGURE 3.2 A two-dimensional rotation of Θ degrees.

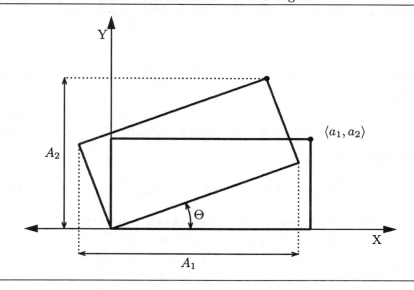

$$= \sum_{k=1}^{n} a \left| \cos \alpha_{ik} \right|$$

$$= a \sum_{k=1}^{n} \left| \cos \alpha_{ik} \right|$$

$$\geq a \sum_{k=1}^{n} \cos^2 \alpha_{ik}$$

$$= a$$

and consequently,

$$\prod_{i=1}^{n} [1 + \mu(A_{i_t})] + \mu(A) - \prod_{i=1}^{n} [\mu(A_{i_t})] \geq \prod_{i=1}^{n} [1 + \mu(A_{i_t})] \geq (1+a)^n . \quad (3.96)$$

Hence, the additivity holds.

The Two–Dimensional Case

Let set A be a rectangle in the standard coordinate system that is shown in Figure 3.2. Since $A = [0, a_1] \times [0, a_2]$ in this system, we have

$$\prod_{i=1}^{2} [1 + \mu(A_{i_t})] + \mu(A) - \prod_{i=1}^{2} [\mu(A_{i_t})] = (1 + a_1)(1 + a_2) \quad (3.97)$$

Now we prove that this is the minimum for all rotations. In the two–dimensional space, any rotation can be represented by the matrix

$$\mathbf{I} = \begin{bmatrix} \cos\theta & -\sin\theta \\ \sin\theta & \cos\theta \end{bmatrix}. \tag{3.98}$$

Fig. 3.2 illustrates a rotated rectangle A and its projections. It is easy to show that

$$\mu(A_1) = a_1 |\cos\theta| + a_2 |\sin\theta| \tag{3.99}$$
$$\mu(A_2) = a_1 |\sin\theta| + a_2 |\cos\theta| . \tag{3.100}$$

Then under the new coordinate system,

$$\prod_{i=1}^{2}[1 + \mu(A_{i_t})] + \mu(A) - \prod_{i=1}^{2}[\mu(A_{i_t})] \tag{3.101}$$

$$= 1 + a_1 |\cos\theta| + a_2 |\sin\theta| + a_1 |\sin\theta| + a_2 |\cos\theta|$$
$$+ (a_1 |\cos\theta| + a_2 |\sin\theta|)(a_1 |\sin\theta| + a_2 |\cos\theta|) + a_1 a_2$$
$$\geq 1 + a_1 \cos^2\theta + a_2 \sin^2\theta + a_1 \sin^2\theta + a_2 \cos^2\theta$$
$$+ (a_1 \cos^2\theta + a_2 \sin^2\theta)(a_1 \sin^2\theta + a_2 \cos^2\theta) + a_1 a_2$$
$$\geq (1 + a_1)(1 + a_2)$$

Therefore, in the two–dimensional case the measure is additive.

The Three–Dimensional Case

To prove (3.83) in the three–dimensional space, we only need to prove that for any rotation of set $A \in C$,

$$\mu(A_1)\mu(A_2) + \mu(A_1)\mu(A_3) + \mu(A_2)\mu(A_3) \geq a_1 a_2 + a_1 a_3 + a_2 a_2 . \tag{3.102}$$

Since the Cartesian product of the projections includes the original set as a subset and both of them are cubes, the area of the surface of the Cartesian projection, which is twice the left hand side of the above inequality, is greater than or equal to the area of the left hand side. Therefore, the inequality holds.

The General n–Dimensional Case

As mentioned previously, it has not been proven yet that the proposed function is additive for sets with unequal edges when $n > 3$. However the additivity of the function has been tested on the computer for thousands of randomly generated examples for each of $n = 4, 5, 6, ..., 10$. In all these examples, additivity of the function has been verified.[1]

---

[1] According to personal communication received during the production of this book, Arthur Ramer proved that HL is additive for any n; the proof was presented at the 1997 IFSA Congress in Prague (June 25-29, 1997) and is included in the Proceedings of the Congress (Volume IV, pp. 268-271).

Generalized Hartley-Like Measures of Nonspecificity

The Hartley-like function is applicable to measuring nonspecificity of convex fuzzy sets defined on $\mathbb{R}^n$ or bodies of evidence defined on $\mathbb{R}^n$ whose focal elements are convex sets. This can readily be accomplished by replacing the Hartley function with the Hartley–like function in the various formulas of this section. Let the U–uncertainty and N–uncertainty based on the Hartley–like function be denoted by UL and NL, respectively. Then, for example, the counterpart of (3.38) is the formula

$$UL(F) = \int_0^{h(F)} HL\left(^\alpha F\right)\, d\alpha + (1 - h(F))\, HL\left(X\right), \qquad (3.103)$$

which is applicable to any possibilistic body of evidence that is derived from a convex fuzzy set (relation) on $\mathbb{R}^n$ for some $n \geq 1$. When $n = 1$ then this formula assumes the simple form

$$UL(F) = \int_0^{h(F)} \ln\left[1 + \mu\left(^\alpha F\right)\right]\, d\alpha + (1 - h(F)) \ln\left(1 + \mu(X)\right). \qquad (3.104)$$

Similarly, the counterpart of (3.54) is the formula

$$NL(m) = \sum_{A \in \mathcal{F}} m(A)\, HL(A), \qquad (3.105)$$

provided that $\mathcal{F}$ is finite.

3.2 Conflict

As indicated in the previous section, the measure of nonspecificity N does not discriminate among probability measures. All probability measures are fully specific: when $\langle \mathcal{F}, m \rangle$ is a probabilistic body of evidence ($\mathcal{F}$ consists of singletons), then $N(m) = 0$. What, then, is actually measured by the Shannon entropy, which is a well established measure of uncertainty in probability theory? Before attempting to answer this question, let us overview key properties of the Shannon entropy.

Shannon Entropy

The Shannon entropy, S, which is applicable only to probability measures, assumes in evidence theory the form

$$S(m) = - \sum_{x \in X} m(\{x\}) \log_2 m(\{x\}). \qquad (3.106)$$

This function, which forms the basis of *classical information theory* [see, e.g., Ash, 1965; Blahut, 1987; Feinstein, 1958; Guiasu, 1977; Kullback, 1959; Reza, 1961], measures the average uncertainty (in bits) associated with the prediction of outcomes in a random experiment; its range is $[0, log_2|X|]$. Clearly, $S(m) = 0$ when $m(\{x\}) = 1$ for some $x \in X$; $S(m) = log_2|X|$ when m defines the uniform probability distribution on X (i.e., $m(\{x\}) = 1/|X|$ for all $x \in X$).

As the name suggests, function S was proposed by Shannon [1948]. It was proven in numerous ways, from several well-justified axiomatic characterizations, that this function is the only sensible measure of uncertainty in probability theory [Aczél and Daróczy, 1975; Mathai and Rathie, 1975].

When m represents a probability distribution then all of the focal sets are singletons. Let us therefore set $p_i = p(x_i) = m(\{x_i\})$ (which is the usual notation for a probability distribution). Then the form of the Shannon entropy is

$$S(p_1, p_2, \ldots, p_n) = - \sum_{i=1}^{n} p_i \log_2 p_i. \tag{3.107}$$

Let $\mathcal{P}$ denote the set of all finite probability distributions. Different subsets of the following requirements, which are universally considered essential for a probabilistic measure of uncertainty and information, are usually taken as the axioms for characterizing the measure:

(S1) Expansibility — When a component with zero probability is added to the probability distribution, the uncertainty should not change; formally,

$$S(p_1, p_2, \ldots, p_n) = S(p_1, p_2, \ldots, p_n, 0) \tag{3.108}$$

for all $\langle p_1, p_2, \ldots, p_n \rangle \in \mathcal{P}$.

(S2) Symmetry — The uncertainty should be invariant with respect to permutations of probabilities of a given probability distribution; formally

$$S(p_1, p_2, \ldots, p_n) = S(\pi(p_1, p_2, \ldots, p_n)) \tag{3.109}$$

for all $\langle p_1, p_2, \ldots, p_n \rangle \in \mathcal{P}$ and for all permutations $\pi(p_1, p_2, \ldots, p_n)$.

(S3) Continuity — Function S should be continuous in all its arguments $p_1, p_2, \ldots, p_n$ ($n \in \mathbb{N}$). This requirement is often replaced with a weaker requirement: $S(p, 1 - p)$ is a continuous function of p in the interval $[0, 1]$.

(S4) Maximum — For each $n \in \mathbb{N}$, the maximum uncertainty should be obtained when all the probabilities are equal to $1/n$. Formally,

$$S(p_1, p_2, \ldots, p_n) \leq S\left(\frac{1}{n}, \frac{1}{n}, \ldots, \frac{1}{n}\right). \tag{3.110}$$

(S5) Subadditivity — The uncertainty of any joint probability distribution should not be greater than the sum of the uncertainties of the corresponding marginal distributions. Formally,

$$S(p_{11}, p_{12}, \ldots, p_{1m}, p_{21}, p_{22}, \ldots, p_{2m}, \cdots, p_{n1}, p_{n2}, \ldots, p_{nm}) \leq \quad (3.111)$$

$$S\left(\sum_{j=1}^{m} p_{1j}, \sum_{j=1}^{m} p_{2j}, \ldots, \sum_{j=1}^{m} p_{nj}\right) + S\left(\sum_{i=1}^{n} p_{i1}, \sum_{i=1}^{n} p_{i2}, \ldots, \sum_{i=1}^{n} p_{im}\right).$$

(S6) Additivity — The uncertainty of any joint probability distribution that is noninteractive should be equal to the sum of the uncertainties of the corresponding marginal distributions. Formally,

$$S(p_1 q_1, p_1 q_2, \ldots, p_1 q_m, p_2 q_1, p_2 q_2, \ldots, p_2 q_m, \cdots, p_n q_1, p_n q_2, \ldots, p_n q_m) =$$
$$S(p_1, p_2, \ldots, p_n) + S(q_1, q_2, \ldots, q_m). \quad (3.112)$$

This requirement is sometimes replaced with a weaker requirement of special additivity for uniform marginal probability distributions with $p_i = 1/n$ and $q_j = 1/m$. Formally,

$$S\left(\tfrac{1}{nm}, \tfrac{1}{nm}, \ldots, \tfrac{1}{nm}\right) = \quad (3.113)$$
$$S\left(\tfrac{1}{n}, \tfrac{1}{n}, \ldots, \tfrac{1}{n}\right) + S\left(\tfrac{1}{m}, \tfrac{1}{m}, \ldots, \tfrac{1}{m}\right).$$

for all $n, m \in \mathbb{N}$. Introducing a convenient function f such that $f(n) = S(\tfrac{1}{n}, \tfrac{1}{n}, \ldots, \tfrac{1}{n})$, then the *weak additivity* can be expressed by the equation

$$f(nm) = f(n) + f(m) \quad (3.114)$$

for all $n, m \in \mathbb{N}$.

(S7) Monotonicity — For probability distributions with equal probabilities $1/n$ ($n \in \mathbb{N}$), the uncertainty should increase with increasing n. Formally,

$$n < m \Rightarrow f(n) < f(m) \quad (3.115)$$

for all $n, m \in \mathbb{N}$, where f denotes the function introduced in (S6).

(S8) Branching — Given a probability distribution on a given finite set X, we require that the total amount of uncertainty should be the same whether it is calculated globally or locally. By locally we mean that the uncertainty is calculated in two stages: the first stage determines the uncertainty of two disjoint subsets of X while the second stage calculates the relative uncertainty within the disjoint subsets. Formally, let $A = \{x_1, x_2, \ldots, x_s\}$ and $B = \{x_{s+1}, x_{s+2}, \ldots, x_n\}$ with $A \cap B = \emptyset$ and $A \cup B = X$. Given a probability distribution on X let

$$p_A = \sum_{i=1}^{s} p_i \quad (3.116a)$$

$$p_B = \sum_{i=s+1}^{n} p_i \qquad (3.116b)$$

denote the probabilities of A and B, respectively. Then the branching axiom can be specified by the equation

$$S(p_1, p_2, \ldots, p_n) = S(p_A, p_B) \qquad (3.117)$$
$$+p_A S\left(\frac{p_1}{p_A}, \frac{p_2}{p_A}, \ldots, \frac{p_s}{p_A}\right)$$
$$+p_B S\left(\frac{p_{s+1}}{p_B}, \frac{p_{s+2}}{p_B}, \ldots, \frac{p_n}{p_B}\right),$$

which should be satisfied for any $\langle p_1, p_2, \ldots, p_n \rangle \in \mathcal{P}$ and for any disjoint subsets A and B that partition X. This requirement, which is also called a grouping requirement, is also presented in alternate forms. For example, one of the weaker forms is given by the formula

$$S(p_1, p_2, p_3) = S(p_1 + p_2, p_3) \qquad (3.118)$$
$$+(p_1 + p_2) S\left(\frac{p_1}{p_1 + p_2}, \frac{p_2}{p_1 + p_2}\right).$$

It matters little which of these forms is adopted since they can be derived from one another.

(S9) Normalization — To ensure (if desirable) that the measurement units of S are bits, it is essential that

$$S\left(\frac{1}{2}, \frac{1}{2}\right) = 1. \qquad (3.119)$$

This axiom must be appropriately modified when other measurement units are preferred.

The listed axioms for a probabilistic measure of information are extensively discussed in the plentiful literature on classical information theory. The following subsets of these axioms are the best known examples of axiomatic characterization of the probabilistic measure of uncertainty:

1. Continuity, weak additivity, monotonicity, branching, and normalization.

2. Expansibility, continuity, maximum, branching, and normalization.

3. Symmetry, continuity, branching, and normalization.

4. Expansibility, symmetry, continuity, subadditivity, additivity, and normalization.

Any of these collections of axioms (as well as some additional collections) is sufficient to characterize the Shannon entropy uniquely. That is, it has been proven that the Shannon entropy is the only function that satisfies any of these sets of axioms. To illustrate in detail this important issue of uniqueness, which gives the Shannon entropy its great significance, we present the uniqueness proof for the first of the listed sets of axioms.

Theorem 3.6 *The Shannon entropy is the unique measure of probabilistic uncertainty, $S : \mathcal{P} \to [0, \infty)$, that satisfies the axioms of continuity, weak additivity (Eq. (3.114)), monotonicity (Eq. (3.115)), branching (Eq. (3.117)), and normalization (Eq. (3.119)).*

Proof. (i) First we prove the proposition $f(n^k) = kf(n)$ for all $n, k \in \mathbb{N}$ by induction on n, where

$$f(n) = S\left(\frac{1}{n}, \frac{1}{n}, \ldots, \frac{1}{n}\right) \tag{3.120}$$

is the same function used in the definition of weak additivity. For $k = 1$ the proposition is trivially true. By the axiom of weak additivity we have

$$f(n^{k+1}) = f(n^k \cdot n) = f(n^k) + f(n). \tag{3.121}$$

Assume the proposition is true for some $k \in \mathbb{N}$. Then,

$$\begin{aligned} f(n^{k+1}) &= f(n^k) + f(n) \\ &= kf(n) + f(n) \\ &= (k+1)f(n), \end{aligned} \tag{3.122}$$

which demonstrates the proposition is true for all $k \in \mathbb{N}$.

(ii) Next we shall demonstrate that $f(n) = \log_2 n$. This proof is identical to that of Theorem 3.1 provided that we replace the Hartley function H with the Shannon function f. Therefore we do not repeat the derivation here. Observe that the proof requires the weak additivity, monotonicity, and normalization.

(iii) We prove now that $S(p, 1 - p) = -p \log_2 p - (1 - p) \log_2(1 - p)$ for rational p. Let $p = r/s$ where $r, s \in \mathbb{N}$. Then

$$\begin{aligned} f(s) &= S\left(\underbrace{\frac{1}{s}, \frac{1}{s}, \ldots, \frac{1}{s}}_{r}, \underbrace{\frac{1}{s}, \frac{1}{s}, \ldots, \frac{1}{s}}_{s-r}\right) \\ &= S\left(\frac{r}{s}, \frac{s-r}{s}\right) + \frac{r}{s} f(r) + \frac{s-r}{s} f(s-r) \end{aligned} \tag{3.123}$$

by the branching axiom. By (ii) and the definition of p we obtain

$$\log_2 s = S(p, 1 - p) + p \log_2 r + (1 - p) \log_2(s - r). \tag{3.124}$$

Solving this equation for $S(p, 1-p)$ results in

$$
\begin{aligned}
S(p, 1-p) &= \log_2 s - p\log_2 r - (1-p)\log_2(s-r) && (3.125)\\
&= p\log_2 s - p\log_2 s + \log_2 s - p\log_2 r - (1-p)\log_2(s-r)\\
&= p\log_2 s - p\log_2 r + (1-p)\log_2 s - (1-p)\log_2(s-r)\\
&= -p\log_2\left(\frac{r}{s}\right) - (1-p)\log_2\left(\frac{s-r}{s}\right)\\
&= -p\log_2 p - (1-p)\log_2(1-p).
\end{aligned}
$$

(iv) We now extend (iii) to the real numbers $p \in [0,1]$ with the help of the continuity axiom. Let p be any number in the unit interval and let p' be a series of rational numbers that approaches p as a limit. Then,

$$
S(p, 1-p) = \lim_{p' \to p} S(p', 1-p') \qquad (3.126)
$$

by the continuity axiom. Moreover,

$$
\begin{aligned}
\lim_{p' \to p} S(p', 1-p') &= \lim_{p' \to p} [-p'\log_2 p' - (1-p')\log_2(1-p')] && (3.127)\\
&= -p\log_2 p - (1-p)\log_2(1-p)
\end{aligned}
$$

since all the functions involved are continuous.

(v) We now conclude the proof by showing that

$$
S(p_1, p_2, \ldots, p_n) = -\sum_{i=1}^{n} p_i \log_2 p_i. \qquad (3.128)
$$

This is accomplished by induction on n. The result is proved in (ii) and (iv) for $n = 1, 2$, respectively. For $n \geq 3$, we may use the branching axiom to obtain

$$
\begin{aligned}
S(p_1, p_2, \ldots, p_n) &= S(p_A, p_n) && (3.129)\\
&\quad + p_A\, S\left(\frac{p_1}{p_A}, \frac{p_2}{p_A}, \ldots, \frac{p_{n-1}}{p_A}\right)\\
&\quad + p_n\, S\left(\frac{p_n}{p_n}\right),
\end{aligned}
$$

where $p_A = \sum_{i=1}^{n-1} p_i$. Since $S(\frac{p_n}{p_n}) = S(1) = 0$ by (ii), we obtain

$$
S(p_1, p_2, \ldots, p_n) = S(p_A, p_n) + p_A\, S\left(\frac{p_1}{p_A}, \frac{p_2}{p_A}, \ldots, \frac{p_{n-1}}{p_A}\right). \qquad (3.130)
$$

By (iv) and assuming the proposition to be true for $n-1$, we may rewrite this equation as

$$
S(p_1, p_2, \ldots, p_n) = S(p_A, p_n) + p_A S\left(\frac{p_1}{p_A}, \frac{p_2}{p_A}, \ldots, \frac{p_{n-1}}{p_A}\right) \qquad (3.131)
$$

$$= -p_A \log_2 p_A - p_n \log_2 p_n - p_A \sum_{i=1}^{n-1} \frac{p_i}{p_A} \log_2 \frac{p_i}{p_A}$$

$$= -p_A \log_2 p_A - p_n \log_2 p_n - \sum_{i=1}^{n-1} p_i \log_2 \frac{p_i}{p_A}$$

$$= -p_A \log_2 p_A - p_n \log_2 p_n$$
$$- \sum_{i=1}^{n-1} p_i \log_2 p_i + \sum_{i=1}^{n-1} p_i \log_2 p_A$$

$$= -p_A \log_2 p_A - p_n \log_2 p_n - \sum_{i=1}^{n-1} p_i \log_2 p_i + p_A \log_2 p_A$$

$$= -\sum_{i=1}^{n} p_i \log_2 p_i .$$

∎

The literature dealing with information theory based on the Shannon entropy is extensive. We do not attempt to give a comprehensive coverage of the theory in this book. However, we briefly overview the most fundamental properties of the Shannon entropy.

First, we present a theorem that plays an important role in classical information theory. This theorem is essential for proving some basic properties of Shannon entropy as well as introducing some additional important concepts of information theory.

Theorem 3.7 *The inequality*

$$-\sum_{i=1}^{n} p_i \log_2 p_i \le -\sum_{i=1}^{n} p_i \log_2 q_i \qquad (3.132)$$

is satisfied for all probability distributions $\langle p_i \mid i \in \mathbb{N}_n \rangle$ and $\langle q_i \mid i \in \mathbb{N}_n \rangle$ and for all $n \in \mathbb{N}$; the equality in (3.132) holds if and only if $p_i = q_i$ for all $i \in \mathbb{N}_n$.

Proof. Consider the function

$$s(p_i, q_i) = p_i \left(\ln p_i - \ln q_i \right) - p_i + q_i \qquad (3.133)$$

for $p_i, q_i \in [0, 1]$. This function is finite, differentiable for all values of p_i and q_i except the pair $p_i = 0$ and $q_i \ne 0$. For each fixed $q_i \ne 0$, the partial derivative of s with respect to p_i is

$$\frac{\partial s(p_i, q_i)}{\partial p_i} = \ln p_i - \ln q_i . \qquad (3.134)$$

That is

$$\frac{\partial s(p_i, q_i)}{\partial p_i} \begin{cases} < 0 & \text{for} \quad p_i < q_i \\ = 0 & \text{for} \quad p_i = q_i \\ > 0 & \text{for} \quad p_i > q_i \end{cases} \qquad (3.135)$$

and, consequently, s is a convex function of p_i, with its minimum at $p_i = q_i$. Hence, for any given i, we have

$$p_i \left(\ln p_i - \ln q_i \right) - p_i + q_i \geq 0 , \qquad (3.136)$$

where the equality holds if and only if $p_i = q_i$. This inequality is also satisfied for $q_i = 0$, since the expression on its left hand side is $+\infty$ if $p_i \neq 0$ and $q_i = 0$, and it is zero if $p_i = 0$ and $q_i = 0$. Taking the sum of this inequality for all $i \in \mathbb{N}_n$, we obtain

$$\sum_{i=1}^{n} \left[p_i \ln p_i - p_i \ln q_i - p_i + q_i \right] \geq 0 , \qquad (3.137)$$

which can be rewritten as

$$\sum_{i=1}^{n} p_i \ln p_i - \sum_{i=1}^{n} p_i \ln q_i - \sum_{i=1}^{n} p_i + \sum_{i=1}^{n} q_i \geq 0 . \qquad (3.138)$$

The last two terms on the left–hand side of this inequality cancel each other out as they both sum up to one. Hence,

$$\sum_{i=1}^{n} p_i \ln p_i - \sum_{i=1}^{n} p_i \ln q_i \geq 0 , \qquad (3.139)$$

which is equivalent to (3.132) when multiplied through by $\frac{1}{\ln 2}$. ∎

This theorem, sometimes referred to as Gibbs' theorem, is quite useful in studying properties of the Shannon entropy. For example, the theorem can be used as follows for proving that the maximum of the Shannon entropy for probability distributions with n elements is $\log_2 n$.

Let $q_i = \frac{1}{n}$ for all $i \in \mathbb{N}_n$. Then, (3.132) yields

$$S(p_i \mid i \in \mathbb{N}_n) = - \sum_{i=1}^{n} p_i \log_2 p_i \leq - \sum_{i=1}^{n} p_i \log_2 \frac{1}{n} \qquad (3.140)$$

and

$$- \sum_{i=1}^{n} p_i \log_2 \frac{1}{n} = - \log_2 \frac{1}{n} \sum_{i=1}^{n} p_i = \log_2 n . \qquad (3.141)$$

Thus, $S(p_i \mid i \in \mathbb{N}_n) \leq \log_2 n$. The upper bound is obtained for $p_i = \frac{1}{n}$ for all $i \in \mathbb{N}_n$.

Let us examine now Shannon entropies of joint, marginal, and conditional probability distributions defined on sets X and Y. In agreement with a common practice in the literature dealing with Shannon entropy, we simplify the notation in the rest of the section by using $S(X)$ instead of $S(p(x) \mid x \in X)$ or $S(p_1, p_2, ..., p_n)$. Furthermore, assuming $x \in X$ and

$y \in Y$, we use symbols $p(x)$ and $p(y)$ to denote marginal probability on sets X and Y, respectively, the symbol $p(x, y)$ for joint probabilities on $X \times Y$, and the symbols $p(x|y)$ and $p(y|x)$ for the corresponding conditional probabilities. In this simplified notation, the meaning of each symbol is uniquely determined by the arguments shown in the parentheses..

Given two sets X and Y, we can recognize three types of entropies:

1. Two *simple entropies* based on marginal probability distributions,

$$S(X) = - \sum_{x \in X} p(x) \log_2 p(x) \qquad (3.142)$$

and

$$S(Y) = - \sum_{y \in Y} p(y) \log_2 p(y) \qquad (3.143)$$

2. A *joint entropy* defined in terms of the joint probability distribution on $X \times Y$,

$$S(X, Y) = - \sum_{\langle x,y \rangle \in X \times Y} p(x, y) \log_2 p(x, y) \qquad (3.144)$$

3. Two *conditional entropies* defined in terms of weighted averages of local conditional entropies as

$$S(X|Y) = - \sum_{y \in Y} p(y) \sum_{x \in X} p(x|y) \log_2 p(x|y) \qquad (3.145)$$

and

$$S(Y|X) = - \sum_{x \in X} p(x) \sum_{y \in Y} p(y|x) \log_2 p(y|x) \qquad (3.146)$$

In addition to these entropies, the function

$$T_S(X, Y) = S(X) + S(Y) - S(X, Y) \qquad (3.147)$$

is often used in the literature as a measure of the strength of relationship (in the probabilistic sense) between elements of set X and Y. The function is called the *information transmission*. It is analogous to the function defined by Eqs. (3.29) for Hartley information: it can be generalized to more than two sets in the same way.

Our next subject is an examination of the relationship among the various types of entropies and the information transmission. The key properties of this relationship are expressed by the next several theorems.

Theorem 3.8

$$S(X|Y) = S(X, Y) - S(Y) \qquad (3.148)$$

Proof.

$$S(X|Y) = -\sum_{y \in Y} p(y) \sum_{x \in X} p(x|y) \log_2 p(x|y) \tag{3.149}$$

$$= -\sum_{y \in Y} p(y) \sum_{x \in X} \frac{p(x,y)}{p(y)} \log_2 \frac{p(x,y)}{p(y)}$$

$$= -\sum_{y \in Y} \sum_{x \in X} p(x,y) \log_2 \frac{p(x,y)}{p(y)}$$

$$= -\sum_{y \in Y} \sum_{x \in X} p(x,y) \log_2 p(x,y) + \sum_{y \in Y} \sum_{x \in X} p(x,y) \log_2 p(y)$$

$$= S(X,Y) + \sum_{y \in Y} \sum_{x \in X} p(x,y) \log_2 p(y)$$

$$= S(X,Y) + \sum_{y \in Y} \log_2 p(y) \sum_{x \in X} p(x,y)$$

$$= S(X,Y) + \sum_{y \in Y} p(y) \log_2 p(y)$$

$$= S(X,Y) - S(Y).$$

∎

The same theorem can obviously be proven for the conditional entropy of Y given X as well:

$$S(Y|X) = S(X,Y) - S(X). \tag{3.150}$$

The theorem can be generalized to more than two sets. The general form, which can be derived from either (3.148) or (3.150) is

$$S(X_1, X_2, ..., X_n) = S(X_1) + S(X_2|X_1) + S(X_3|X_1, X_2) \tag{3.151}$$
$$+ \cdots + S(X_n|X_1, X_2, ..., X_{n-1}).$$

This equation is valid for any permutation of the sets involved.

Theorem 3.9

$$S(X,Y) \le S(X) + S(Y) \tag{3.152}$$

Proof.

$$S(X) = -\sum_{x \in X} p(x) \log_2 p(x) \tag{3.153}$$

$$= -\sum_{x \in X} \sum_{y \in Y} p(x,y) \log_2 \sum_{y \in Y} p(x,y)$$

$$S(Y) = -\sum_{y \in Y} p(y) \log_2 p(y) \tag{3.154}$$

$$= -\sum_{y \in Y} \sum_{x \in X} p(x,y) \log_2 \sum_{x \in X} p(x,y)$$

$$S(X) + S(Y) = -\sum_{x \in X} \sum_{y \in Y} p(x,y) \left[\log_2 \sum_{y \in Y} p(x,y) \right. \tag{3.155}$$

$$\left. + \log_2 \sum_{x \in X} p(x,y) \right]$$

$$= -\sum_{\langle x,y \rangle \in X \times Y} p(x,y) \left[\log_2 \ p(x) + \log_2 \ p(y) \right]$$

$$= -\sum_{\langle x,y \rangle \in X \times Y} p(x,y) \log_2 \left[p(x) \cdot p(y) \right] .$$

By Gibbs' theorem we have

$$S(X,Y) = -\sum_{\langle x,y \rangle \in X \times Y} p(x,y) \log_2 p(x,y) \tag{3.156}$$

$$\leq -\sum_{\langle x,y \rangle \in X \times Y} p(x,y) \log_2 \left[p(x) \cdot p(y) \right] = S(X) + S(Y) .$$

Hence $S(X,Y) \leq S(X) + S(Y)$; and furthermore (again by Gibbs' theorem), the equality holds if and only if

$$p(x,y) = p(x) \cdot p(y) , \tag{3.157}$$

that is, only if the sets X and Y are noninteractive in the probabilistic sense. ∎

Theorem 3.9 can easily be generalized to more than two sets. Its general form is

$$S(X_1, X_2, ..., X_n) \leq \sum_{i=1}^{n} S(X_i) \tag{3.158}$$

which holds for every $n \in \mathbb{N}$.

Theorem 3.10

$$S(X) \geq S(X|Y) \tag{3.159}$$

Proof. From Theorem 3.8

$$S(X|Y) = S(X,Y) - S(Y) \tag{3.160}$$

and from Theorem 3.9

$$S(X,Y) \leq S(X) + S(Y) . \tag{3.161}$$

Hence,

$$S(X|Y) + S(Y) \leq S(X) + S(Y) \tag{3.162}$$

and the inequality

$$S(X|Y) \leq S(X) \tag{3.163}$$

follows immediately.   ■

Exchanging X and Y in Theorem 3.10, we obtain

$$S(Y) \geq S(Y|X).$$ (3.164)

Additional equations expressing the relationships among the various entropies and the information transmission can be obtained by simple formula manipulations with the aid of key properties in Theorems 3.7 through 3.10. For example, when we substitute for $S(X,Y)$ from Eq. (3.148) into Eq. (3.147), we obtain

$$T_S(X,Y) = S(X) - S(X|Y);$$ (3.165)

similarly, by substituting Eq. (3.150) into Eq. (3.147), we obtain

$$T_S(X,Y) = S(Y) - S(Y|X).$$ (3.166)

By comparing Eqs. (3.165) and (3.166), we also obtain

$$S(X) - S(Y) = S(X|Y) - S(Y|X).$$ (3.167)

One way of getting insight into the type of uncertainty measured by the Shannon entropy is to rewrite (3.106) in the form

$$S(m) = - \sum_{x \in X} m(\{x\}) \log_2 \left[1 - \sum_{y \neq x} m(\{y\}) \right].$$ (3.168)

The term

$$Con(\{x\}) = \sum_{y \neq x} m(\{y\}).$$ (3.169)

in Eq. (3.168) represents the total evidential claim pertaining to focal elements that are different from the focal element $\{x\}$. That is, $Con(\{x\})$ expresses the sum of all evidential claims that fully conflict with the one focusing on $\{x\}$. Clearly, $Con(\{x\}) \in [0,1]$ for each $x \in X$. The function $-\log_2[1 - Con(\{x\})]$, which is employed in Eq. (3.168), is monotonic increasing with $Con(\{x\})$ and extends its range from $[0,1]$ to $[0,\infty)$. The choice of the logarithmic function is a result of the axiomatic requirement that the joint uncertainty of several independent random variables be equal to the sum of their individual uncertainties [Aczel and Daroczy, 1975; Klir and Folger, 1988; Renyi, 1970].

It follows from these facts and from the form of Eq. (3.168) that the Shannon entropy is the mean (expected) value of the *conflict* among evidential claims within a given probabilistic body of evidence.

One important aspect of Shannon entropy remains to be discussed. It is connected with its restriction to finite sets. Is this restriction necessary? It seems that the formula

$$B(q(x)|x \in [a, b]) = -\int_a^b q(x) \log_2 q(x)\, dx, \qquad (3.170)$$

where q denotes a probability density function on the real interval $[a, b]$, is analogous to Eq. (3.107) for Shannon entropy and could thus be viewed as an extension of Shannon entropy to the domain of real numbers. Moreover, function B defined by Eq. (3.170) is usually referred to as the *Boltzmann entropy*. Although the analogy between the two functions is suggestive, the following question cannot be avoided: is the Boltzmann entropy a genuine extension of the Shannon entropy? To answer this nontrivial question, we must establish a connection between the two functions.

Let q be a probability density function on the interval $[a, b]$ of real numbers. That is, $q(x) \geq 0$ for all $x \in [a, b]$ and,

$$\int_a^b q(x)\, dx = 1. \qquad (3.171)$$

Consider a sequence of probability distributions ${}^n\mathbf{p} = ({}^np_1, {}^np_2, \ldots, {}^np_n)$ such that

$$ {}^np_i = \int_{x_{i-1}}^{x_i} q(x)\, dx \qquad (3.172)$$

for every $n \in \mathbb{N}_n$, where

$$x_i = a + i\,\frac{b - a}{n} \qquad (3.173)$$

for each $i \in \mathbb{N}_n$, and $x_0 = a$ by convention. For convenience, let

$$\Delta_n = \frac{b - a}{n} \qquad (3.174)$$

so that

$$x_i = a + i\,\Delta_n. \qquad (3.175)$$

For each probability distribution ${}^n\mathbf{p} = ({}^np_1, {}^np_2, \ldots, {}^np_n)$, let ${}^n\mathbf{d}(x)$ denote a probability density function on $[a, b]$ such that

$$ {}^n\mathbf{d}(x) = ({}^nd_i(x) \,|\, i \in \mathbb{N}_n), \qquad (3.176)$$

where

$$ {}^nd_i(x) = \frac{{}^np_i}{\Delta_n} \text{ for } x \in [x_{i-1}, x_i) \qquad (3.177)$$

and for all $i \in \mathbb{N}_n$. Then due to continuity of $q(x)$, the sequence $({}^n\mathbf{d}(x) \,|\, n \in \mathbb{N})$ converges to $q(x)$ uniformly on $[a, b]$.

Given the probability distribution ${}^n\mathbf{p}$ for some $n \in \mathbb{N}$, its Shannon entropy is

$$S({}^n\mathbf{p}) = -\sum_{i=1}^n {}^np_i \log_2 {}^np_i \qquad (3.178)$$

or, using the introduced probability density function $^n\mathbf{d}$,

$$S(^n\mathbf{p}) = -\sum_{i=1}^{n} {}^nd_i(x)\,\Delta_n\,\log_2[\,^nd_i(x)\,\Delta_n]. \tag{3.179}$$

This equation can be modified as follows:

$$S(^n\mathbf{p}) = -\quad \sum_{i=1}^{n} {}^nd_i(x)\,\Delta_n\,\log_2\,{}^nd_i(x) \tag{3.180}$$

$$-\sum_{i=1}^{n} {}^nd_i(x)\,\Delta_n\,\log_2\Delta_n$$

$$= -\quad \sum_{i=1}^{n}[\,^nd_i(x)\,\log_2\,{}^nd_i(x)]\,\Delta_n - \log_2\Delta_n\sum_{i=1}^{n} {}^np_i\,.$$

Since probabilities ^np_i of the distribution $^n\mathbf{p}$ must add to one, and by the definition of Δ_n, we obtain

$$S(^n\mathbf{p}) = -\sum_{i=1}^{n}[\,^nd_i(x)\,\log_2\,{}^nd_i(x)]\,\Delta_n + \log_2\frac{n}{b-a}. \tag{3.181}$$

When $n \to \infty$(or $\Delta_n \to 0$), we have

$$\lim_{n\to\infty} -\sum_{i=1}^{n}[\,^nd_i(x)\,\log_2\,{}^nd_i(x)]\,\Delta_n = -\int_a^b q(x)\,\log_2 q(x)\,dx \tag{3.182}$$

according to the introduced relation among $^n\mathbf{p}$, $q(x)$, and $^nd_i(x)$, in particular Eqs. (3.172) and (3.177). Equation (3.181) can thus be written for $n \to \infty$ as

$$\lim_{n\to\infty} S(^n\mathbf{p}) = B(q(x)) + \lim_{n\to\infty}\frac{n}{b-a}. \tag{3.183}$$

The last term in this equation clearly diverges. This means that the Boltzmann entropy *is not* a limit of the Shannon entropy for $n \to \infty$ and, consequently, it is not a measure of uncertainty and information.

The discrepancy between the Shannon and Boltzmann entropies can be reconciled in a modified form of the Boltzmann entropy,

$$\hat{B}[f(x),g(x)|x \in [a,b]] = \int_a^b f(x)\,\log_2\frac{f(x)}{g(x)}\,dx\,, \tag{3.184}$$

which involves two probability density functions, $f(x)$ and $g(x)$, defined on $[a,b]$. Its finite counterpart is

$$\hat{S}\,[p(x),q(x) \mid x \in X] = \sum_{x\in X} p(x)\,\log_2\frac{p(x)}{q(x)}. \tag{3.185}$$

This function is known in information theory as the *Shannon cross-entropy* or *directed divergence*. Function $\hat{S}$ measures uncertainty in relative rather than absolute terms.

When $f(x)$ in Eq. (3.184) is replaced with a density function, $f(x, y)$, of a joint probability distribution on $X \times Y$, and $g(x)$ is replaced with the product of density functions of marginal distributions on X and Y, $f(x) \cdot f(y)$, $\hat{B}$ becomes the continuous counterpart to the information transmission given by Eq. (3.147). This means that the continuous counterpart, T_B, of the information transmission can be expressed as

$$T_B[f(x, y), f(x) \cdot f(y) \mid x \in [a, b], y \in [c, d]] = \qquad (3.186)$$
$$\int_a^b \int_c^d f(x, y) \log_2 \frac{f(x, y)}{f(x) \cdot f(y)} \ dx \ dy.$$

Entropy-Like Measure in Evidence Theory

What form the generalized counterpart of the Shannon entropy in evidence theory should take? The answer is by no means obvious, as exhibited by several proposed candidates for the entropy-like measure in evidence theory. Let us introduce and critically examine each of them.

Two of the candidates were proposed in the early 1980s. One of them is function E defined by the formula

$$E(m) = - \sum_{A \in \mathcal{F}} m(A) \log_2 Pl(A), \qquad (3.187)$$

which is usually called a measure of *dissonance;* this function was proposed by Yager [1983]. The other one is function C defined by the formula

$$C(m) = - \sum_{A \in \mathcal{F}} m(A) \log_2 Bel(A) \qquad (3.188)$$

which is referred to as a measure of *confusion;* it was proposed by Höhle [1982]. It is obvious that both of these functions collapse into the Shannon entropy when m defines a probability measure.

To decide if either of the two functions is an appropriate generalization of the Shannon entropy in evidence theory, we have to determine what do these functions actually measure.

From Eq. (2.52) and the general property of basic probability assignments (satisfied for every $A \in \mathcal{P}(X)$),

$$\sum_{A \cap B = \emptyset} m(B) + \sum_{A \cap B \neq \emptyset} m(B) = 1, \qquad (3.189)$$

we obtain

$$E(m) = - \sum_{A \in \mathcal{F}} m(A) \log_2 \left(1 - \sum_{A \cap B = \emptyset} m(B) \right). \qquad (3.190)$$

The term

$$K(A) = \sum_{A \cap B = \emptyset} m(B) \qquad (3.191)$$

in Eq. (3.190) represents the total evidential claim pertaining to focal elements that are disjoint with the set A. That is, $K(A)$ expresses the sum of all evidential claims that fully conflict with the one focusing on the set A. Clearly, $K(A) \in [0, 1]$. The function

$$-\log_2[1 - K(A)], \qquad (3.192)$$

which is employed in Eq. (3.190), is monotonic increasing with $K(A)$ and extends its range from $[0, 1]$ to $[0, \infty)$. The choice of the logarithmic function is motivated in the same way as in the classical case of the Shannon entropy.

It follows from these facts and the form of Eq. (3.190) that $E(m)$ is the mean (expected) value of the conflict among evidential claims within a given body of evidence $\langle \mathcal{F}, m \rangle$; it measures the conflict in bits and its range is $[0, log_2|X|]$.

Function E is not fully satisfactory since we feel intuitively that $m(B)$ conflicts with $m(A)$ not only when $B \cap A = \emptyset$. This broader view of conflict is expressed by the measure of confusion C given by Eq. (3.188). Let us demonstrate this fact.

From Eq. (2.51) and the general property of basic assignments (satisfied for every $A \in \mathcal{P}$),

$$\sum_{B \subseteq A} m(B) + \sum_{B \not\subseteq A} m(B) = 1 \qquad (3.193)$$

we get

$$C(m) = - \sum_{A \in \mathcal{F}} m(A) \log_2 \left(1 - \sum_{B \not\subseteq A} m(B) \right) \qquad (3.194)$$

The term

$$L(A) = \sum_{B \not\subseteq A} m(B) \qquad (3.195)$$

in Eq. (3.194) expresses the sum of all evidential claims that conflict with the one focusing on the set A according to the following broader view of conflict: $m(B)$ conflicts with $m(A)$ whenever $B \not\subseteq A$. The reason for using the function

$$-\log_2[1 - L(A)] \qquad (3.196)$$

instead of $L(A)$ in Eq. (3.194) is the same as already explained in the context of function E. The conclusion is that $C(m)$ is the mean (expected) value of the conflict, viewed in the broader sense, among evidential claims within a given body of evidence $\langle \mathcal{F}, m \rangle$.

Function C is also not fully satisfactory as a measure of conflicting evidential claims within a body of evidence, but for a different reason than function E. Although it employs the broader, and more satisfactory, view of conflict, it does not properly scale each particular conflict of $m(B)$ with respect to $m(A)$ according to the degree of violation of the subsethood relation $B \subseteq A$. It is clear that the more this subsethood relation is violated the greater the conflict. In addition, function C has also some undesirable mathematical properties. For example, its maximum is greater than $log_2|X|$.

To overcome the deficiencies of functions E and C as adequate measures of conflict in evidence theory, a new function, D, was proposed by Klir and Ramer [1990]:

$$D(m) = -\sum_{A \in \mathcal{F}} m(A) \log_2 \left(1 - \sum_{B \in \mathcal{F}} m(B) \frac{|B - A|}{|B|} \right) \qquad (3.197)$$

Observe that the term

$$Con(A) = \sum_{B \in \mathcal{F}} m(B) \frac{|B - A|}{|B|} \qquad (3.198)$$

in Eq. (3.197) expresses the sum of individual conflicts of evidential claims with respect to a particular set A, each of which is properly scaled by the degree to which the subsethood $B \subseteq A$ is violated. This conforms to the intuitive idea of conflict that emerged from the critical re-examination of functions E and C. Let function Con, whose application to probability measures is given by Eq. (3.169), be called a *conflict*. Clearly, $Con(A) \in [0, 1]$ and, furthermore,

$$K(A) \leq Con(A) \leq L(A) \qquad (3.199)$$

The reason for using the function

$$-\log_2[1 - Con(A)] \qquad (3.200)$$

instead of Con in Eq.(3.197) is exactly the same as previously explained in the context of function E. This monotonic transformation extends the range of $Con(A)$ from $[0, 1]$ to $[0, \infty)$.

Function D, which is called a measure of *discord,* is clearly a measure of the mean conflict (expressed by the logarithmic transformation of function Con) among evidential claims within each given body of evidence. It follows immediately from (3.199) that

$$E(m) \leq D(m) \leq C(m). \qquad (3.201)$$

Observe that $|B - A| = |B| - |A \cap B|$ and, consequently, Eq. (3.197) can be rewritten as

$$D(m) = -\sum_{A \in \mathcal{F}} m(A) \log_2 \sum_{B \in \mathcal{F}} m(B) \frac{|A \cap B|}{|B|} \qquad (3.202)$$

It is obvious that

$$Bel(A) \leq \sum_{B \in \mathcal{F}} m(B)\frac{|A \cap B|}{|B|} \leq Pl(A) \tag{3.203}$$

Function D is applicable equally well to the fuzzified evidence theory provided that the cardinality of a fuzzy sets is defined by Eq. (2.16) and the set intersection is defined by the minimum operator.

Although function D is intuitively more appealing than functions E and C, further examination revealed a conceptual defect in it [Klir and Parviz, 1992a]. To explain the defect, let sets A and B in Eq. (3.198) be such that $A \subset B$. Then, according to function Con, the claim $m(B)$ is taken to be in conflict with the claim $m(A)$ to the degree $|B - A|/|B|$. This, however, should not be the case: the claim focusing on B is implied by the claim focusing on A (since $A \subset B$) and, hence, $m(B)$ should not be viewed in this case as contributing to the conflict with $m(A)$.

Consider, as an example, incomplete information regarding the age of a person, say Joe. Assume that the information is expressed by two evidential claims pertaining to the age of Joe: "Joe is between 15 and 17 years old" with degree $m(A)$, where $A = [15, 17]$, and "Joe is a teenager" with degree $m(B)$, where $B = [13, 19]$. Clearly, the weaker second claim does not conflict with the stronger first claim.

Assume now that $A \supset B$. In this case, the situation is inverted: the claim focusing on B is not implied by the claim focusing on A and, consequently, $m(B)$ does conflict with $m(A)$ to a degree proportional to number of elements in A that are not covered by B. This conflict is not captured by function Con since $|B - A| = 0$ in this case.

It follows from these observations that the total conflict of evidential claims within a body of evidence $\langle \mathcal{F}, m \rangle$ with respect to a particular claim $m(A)$ should be expressed by function

$$CON(A) = \sum_{B \in \mathcal{F}} m(B)\frac{|A - B|}{|A|} \tag{3.204}$$

rather than function Con given by Eq. (3.198). Replacing $Con(A)$ in Eq. (3.197) with $CON(A)$, we obtain a new function, which is better justified as a measure of conflict in evidence theory than function D. This new function, which is called *strife* and denoted by ST, is defined by the formula

$$ST(m) = -\sum_{A \in \mathcal{F}} m(A) \log_2 \left(1 - \sum_{B \in \mathcal{F}} m(B)\frac{|A - B|}{|A|}\right) \tag{3.205}$$

It is trivial to convert this form into a simpler one,

$$ST(m) = -\sum_{A \in \mathcal{F}} m(A) \log_2 \sum_{B \in \mathcal{F}} m(B)\frac{|A \cap B|}{|A|} \tag{3.206}$$

where the term $|A \cap B|/|A|$ expresses the degree of subsethood of set A in set B. Eq. (3.206) can also be rewritten as

$$ST(m) = N(m) - \sum_{A \in \mathcal{F}} m(A) \log_2 \sum_{B \in \mathcal{F}} m(B)|A \cap B| \qquad (3.207)$$

where $N(m)$ is the nonspecificity measure given by Eq. (3.51). Furthermore, introducing

$$Z(m) = \sum_{A \in \mathcal{F}} m(A) \log_2 \sum_{B \in \mathcal{F}} m(B)|A \cap B| \qquad (3.208)$$

we have

$$ST(m) = N(m) - Z(m) \qquad (3.209)$$

As shown by Klir and Yuan [1993], the distinction between strife and discord reflects the distinction between disjunctive and conjunctive set valued statements, respectively. The latter distinction was introduced and studied by Yager [1984, 1987, 1988]. Let us describe the connection between the two distinctions.

In the following, we deal with statements of the form "x is A," where A is a subset of a given universal set X and $x \in X$. We assume the framework of evidence theory, in which the evidence supporting this proposition is expressed by the value $m(A)$ of the basic probability assignment. The statement may be interpreted either as a *disjunctive set valued statement* or a *conjunctive set–valued statement.*

A statement "x is A" is disjunctive if it means that x conforms to one of the elements in A. For example "Mary is a teenager" is disjunctive because it means that the real age of Mary conforms to one of the values in the set $\{13, 14, 15, 16, 17, 18, 19\}$. Similarly "John arrived between 10:00 and 11:00 a.m." is disjunctive because it means that the real John's arrival time was one value in the time interval between 10:00 and 11:00 a.m.

A statement "x is A" is conjunctive if it means that x conforms to all of the elements in A. For example the statement "The compound consists of iron, copper, and aluminium" is conjunctive because it means that the compound in question conforms to all the elements in the set $\{$iron, copper, aluminium$\}$. Similarly, "John was in the doctor's office from 10:00 to 11:00 a.m." is conjunctive because it means that John was in the doctor's office not only at one time during the time interval, but all the time instances during the time interval between 10:00 and 11:00 a.m.

Let S_A and S_B denote, respectively, the statements "x is A" and "x is B." Assume that $A \subset B$ and the statements are disjunctive. Then, clearly, S_A implies S_B and, consequently, S_B does not conflict with S_A while S_A does conflict with S_B. For example, the statement S_B: "Mary is a teenager" does not conflict with the statement S_A: "Mary is fifteen or sixteen," while S_A conflicts with S_B.

Let S_A and S_B be conjunctive and assume again that $A \subset B$. Then, clearly, S_B implies S_A and, consequently, S_A does not conflict with S_B

FIGURE 3.3 Maximum values of possibilistic discord and possibilistic strife for $n = 2, 3, \ldots, 22$.

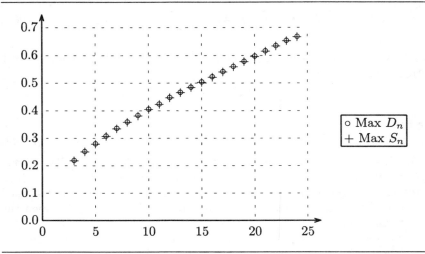

while S_B does conflict with S_A. For example, the statement S_B: "Steel is a compound of iron, carbon, and nickel" does conflict with the statement S_A: "Steel is a compound of iron and carbon," while S_A does not conflicts with S_B in this case.

This examination shows clearly that the measure of strife ST (Eq. (3.206)) expresses the conflict among disjunctive statements, while the measure of discord D (Eq. (3.202)) expresses the conflict among conjunctive statements. Since evidence theory and possibility theory deal usually with disjunctive statements, this reinforces the arguments made by Klir and Parviz [1992b] that the measure of strife is a better justified measure of conflict (or entropy-like measure) in evidence theory and possibility theory.

It is reasonable to conclude that function ST is well justified on intuitive grounds as a measure of conflict among evidential claims in evidence theory when disjunctive statements are employed. Similarly function D is a well justified measure of conflict in evidence theory when conjunctive statements are employed. Both functions also possess some desirable mathematical properties: they are additive, their measurement units are bits, they becomes the Shannon entropy for probability measures, and their range is $[0, log_2|X|]$, where $ST(m) = 0$ when $m(\{x\}) = 1$ for some $x \in X$ and $ST(m) = log_2|X|$ when $m(\{x\}) = \frac{1}{|X|}$ for all $x \in X$ or when $m(X) = 1$. Unfortunately, they are not subadditive [Ramer and Klir, 1993; Vejnarova and Klir, 1993].

Conflict in Possibility Theory

Employing ordered possibility distributions $1 = r_1 \geq r_2 \geq ... \geq r_n$, it is easy to derive the following forms of discord and strife

$$D(\mathbf{r}) = \sum_{i=1}^{n-1} (r_i - r_{i+1}) \log_2 \left[1 - i \sum_{j=i+1}^{n} \frac{r_j}{j(j-1)} \right] \qquad (3.210)$$

and

$$ST(\mathbf{r}) = \sum_{i=2}^{n} (r_i - r_{i+1}) \log_2 \frac{i}{\sum_{j=1}^{i} r_j} \qquad (3.211)$$

Possibilistic discord was thoroughly analyzed by Geer and Klir [1991]. They established that the maximum value of possibilistic discord is bounded from above by $\log_2 e \approx 1.443$ and that the actual upper bound (for $n \to \infty$) is approximately 0.892. It turn out that the possibilistic strife has exactly the same properties. Moreover, the maximum values of possibilistic strife depend on n in exactly the same way as the maximum values of possibilistic discord (Fig. (3.3)). However, the possibility distributions for which the maximum are obtained (one for each value of n) are different for the two functions [Geer and Klir, 1991; Klir, 1993].

We may conclude from these results that the measures of both discord and strife, whose range in evidence theory is $[0, log_2|X|]$, are severely constrained within the domain of possibility theory. We may say that possibility theory is almost conflict-free. For large bodies of evidence, at least, these measures can be considered negligible when compared with the other type of uncertainty – nonspecificity. Neglecting strife (or discord), when justifiable, may substantially reduce computational complexity in dealing with large possibilistic bodies of evidence.

3.3 Aggregate Uncertainty in Evidence Theory

Let AU denote a function by which the *aggregate uncertainty ingrained* in any given body of evidence (expressed in terms of evidence theory) can be measured. This function is supposed to capture, in an aggregate fashion, both nonspecificity and conflict—the two types of uncertainty that coexist in evidence theory. The function AU may be expressed in terms of belief measures, plausibility measures, or basic probability assignments. Choosing, for example, belief measures, the function has the following form

$$AU : \mathcal{B} \to [0, \infty), \qquad (3.212)$$

where $\mathcal{B}$ is the set of all belief measures. Since there are one-to-one mappings between corresponding belief measures, plausibility measures and basic probability assignments, as expressed by Eqs. (2.47) and (2.51)–(2.53), the domain of function AU expressed in this form may be reinterpreted in terms of the corresponding plausibility measures or basic probability assignments.

To qualify as a meaningful measure of aggregate uncertainty in evidence theory, function AU must satisfy certain requirements that are generally considered essential on intuitive grounds. These are the following requirements:

(**AU1**) **Probability consistency** — Whenever Bel defines a probability measure (i.e., all focal subsets are singletons), AU assumes the form of the Shannon entropy

$$AU(\text{Bel}) = -\sum_{x \in X} \text{Bel}(\{x\}) \log_2 \text{Bel}(\{x\}).\qquad(3.213)$$

(**AU2**) **Set consistency** — Whenever Bel focuses on a single set (i.e., $m(A) = 1$ for some $A \subseteq X$), AU assumes the form of the Hartley measure

$$AU(\text{Bel}) = \log_2|A|.\qquad(3.214)$$

(**AU3**) **Range** — The range of AU is $[0, log_2|X|]$ when Bel is defined on $\mathcal{P}(X)$ and AU is measured in bits.

(**AU4**) **Subadditivity** — If Bel is an arbitrary joint belief function on $X \times Y$ and the associated marginal belief functions are Bel_X and Bel_Y, then
$$AU(\text{Bel}) \leq AU(\text{Bel}_X) + AU(\text{Bel}_Y).\qquad(3.215)$$

(**AU5**) **Additivity** — If Bel is a joint belief function on $X \times Y$ and the marginal belief functions Bel_X and Bel_Y are noninteractive, then

$$AU(\text{Bel}) = AU(\text{Bel}_X) + AU(\text{Bel}_Y).\qquad(3.216)$$

The quest to find a measure of the aggregate uncertainty in evidence theory has been long and difficult. An early idea was to aggregate nonspecificity and conflict by adding their measures. This idea was first pursued by Lamata and Moral [1988], who suggested to add functions N, defined by Eq. (3.51), and function E, defined by Eq. (3.187). Later the sums $N + D$ and $N + ST$ were explored by Klir and Ramer [1990], Vejnarova and Klir [1993] , and Ramer and Klir [1993]. Unfortunely, all of these prospective measures of aggregate uncertainty were shown to violate the requirement of subadditivity (AU4). Hence, their justification is questionable.

A measure of aggregate uncertainty in evidence theory that satisfies all of the requirements (AU1)–(AU5) was conceived by several authors in the early 1990s [Maeda and Ichihashi, 1993; Maeda, Nguyen, and Ichihashi, 1993; Harmanec and Klir, 1994]. The measure is defined as follows.

Given a belief measure Bel on the power set of a finite set X, the aggregate uncertainty associated with Bel is measured by the function

$$AU(\text{Bel}) = \max_{\mathcal{P}_{\text{Bel}}} \left[-\sum_{x \in X} p_x \log_2 p_x \right] \qquad (3.217)$$

where the maximum is taken over all probability distributions that are consistent with the given belief measure. Thus, $\mathcal{P}_{\text{Bel}}$ in (3.217) consists of all probability distributions $\langle p_x \mid x \in X \rangle$ that satisfy the constraints

(a) $p_x \in [0,1]$ for all $x \in X$ and $\sum_{x \in X} p_x = 1$.

(b) $\text{Bel}(A) \leq \sum_{x \in A} p_x \leq 1 - \text{Bel}(\bar{A})$ for all $A \subseteq X$.

The following five theorems, which were proven by Harmanec and Klir [1994], establish that the function AU defined by Eq. (3.217) satisfies all of the requirements (AU1)–(AU5). It is thus a well–justified measure of aggregate uncertainty in evidence theory. Although its uniqueness is still an open problem, it was proven by Harmanec [1995] that the function AU is the smallest measure of aggregate uncertainty in evidence theory among all other measures (if they exist).

Theorem 3.11 *The measure AU is probability consistent.*

Proof. When Bel is a probability measure, all focal elements are singletons and this implies that $\text{Bel}(\{x\}) = p_x = \text{Pl}(\{x\})$ for all $x \in X$. Hence the maximum is taken over a single probability distribution $\langle p_x \mid x \in X \rangle$ and $AU(\text{Bel})$ is equal to the Shannon entropy of this distribution. ∎

Theorem 3.12 *The measure AU is set consistent.*

Proof. Let m denote the basic probability assignment corresponding to function Bel in (2.53). By assumption of set consistency $m(A) = 1$ for some $A \subseteq X$ and this implies that $m(B) = 0$ for all $B \neq A$ including $B \subset A$. This means that every probability distribution that sums to one for elements x in A and is zero for all x not in A is consistent with Bel. It is well known that the uniform probability distribution maximizes the entropy function and, hence, the uniform probability distribution on A will maximize AU. That is,

$$AU(\text{Bel}) = -\sum_{x \in A} \frac{1}{|A|} \log_2 \frac{1}{|A|} = \log_2 |A|. \qquad (3.218)$$

∎

Theorem 3.13 *The measure AU has range $[0, \log_2|X|]$.*

Proof. Since $[0, log_2|X|]$ is the range of the Shannon entropy for any probability distribution on X the measure AU cannot be outside these bounds.

∎

Theorem 3.14 *The measure AU is subadditive.*

Proof. Let Bel be a function on $X \times Y$ and let $\langle \hat{p}_{xy} \mid \langle x, y \rangle \in X \times Y \rangle$ denote a probability distribution for which

$$AU(\text{Bel}) = -\sum_{x \in X} \sum_{y \in Y} \hat{p}_{xy} \log_2 \hat{p}_{xy} \qquad (3.219)$$

and

$$\text{Bel}(A) \leq \sum_{\langle x,y \rangle \in A} \hat{p}_{xy} \qquad (3.220)$$

for all $A \subseteq X \times Y$ (this must be true for $\hat{p}_{xy}$ to be consistent with Bel). Furthermore, let

$$\hat{p}_x = \sum_{y \in Y} \hat{p}_{xy} \quad \text{and} \quad \hat{p}_y = \sum_{x \in X} \hat{p}_{xy} \qquad (3.221)$$

Using Gibbs' inequality (3.132), we have

$$-\sum_{x \in X} \sum_{y \in Y} \hat{p}_{xy} \log_2 \hat{p}_{xy} \quad \leq \quad -\sum_{x \in X} \sum_{y \in Y} \hat{p}_{xy} \log_2 (\hat{p}_x \cdot \hat{p}_y)$$

$$= \quad -\sum_{x \in X} \hat{p}_x \log_2 \hat{p}_x - \sum_{y \in Y} \hat{p}_y \log_2 \hat{p}_y \quad .$$

Observe that, for all $A \subseteq X$

$$\text{Bel}_X(A) = \text{Bel}(A \times Y) \leq \sum_{x \in A} \sum_{y \in Y} \hat{p}_{xy} = \sum_{x \in A} \hat{p}_x \qquad (3.222)$$

and, analogously, for all $B \subseteq Y$,

$$\text{Bel}_Y(B) \leq \sum_{y \in B} \hat{p}_y \ . \qquad (3.223)$$

Considering all these facts we, conclude

$$AU(\text{Bel}) \quad = \quad -\sum_{x \in X} \sum_{y \in Y} \hat{p}_{xy} \log_2 \hat{p}_{xy} \qquad (3.224)$$

$$\leq \quad -\sum_{x \in X} \hat{p}_x \log_2 \hat{p}_x - \sum_{y \in Y} \hat{p}_y \log_2 \hat{p}_y$$

$$\leq \quad AU(\text{Bel}_X) + AU(\text{Bel}_Y) \ .$$

∎

Theorem 3.15 *The measure AU is additive.*

Proof. By subadditivity we know that $AU(\mathrm{Bel}) \leq AU(\mathrm{Bel}_X) + AU(\mathrm{Bel}_Y)$. When Bel is noninteractive we need to prove the reverse, $AU(\mathrm{Bel}) \geq AU(\mathrm{Bel}_X) + AU(\mathrm{Bel}_Y)$, to conclude that the two quantities must be equal and that AU is additive.

Let Bel be a noninteractive belief function. Let $\langle \hat{p}_x \mid x \in X \rangle$ be the probability distribution for which

$$AU(\mathrm{Bel}_X) = - \sum_{x \in X} \hat{p}_x \log_2 \hat{p}_x \qquad (3.225)$$

and

$$\mathrm{Bel}(A) \leq \sum_{x \in A} \hat{p}_x \qquad (3.226)$$

for all $A \subseteq X$; similarly, let $\langle \hat{p}_y \mid y \in Y \rangle$ be the probability distribution for which

$$AU(\mathrm{Bel}_Y) = - \sum_{y \in Y} \hat{p}_y \log_2 \hat{p}_y \qquad (3.227)$$

and

$$\mathrm{Bel}_Y(B) \leq \sum_{y \in B} \hat{p}_y \quad . \qquad (3.228)$$

for all $B \subseteq Y$. Define $\hat{p}_{xy} = \hat{p}_x \cdot \hat{p}_y$ for all $\langle x, y \rangle \in X \times Y$. Clearly $\hat{p}_{xy}$ is a probability distribution on $X \times Y$. Morover, for all $C \subseteq X \times Y$,

$$
\begin{aligned}
\sum_{\langle x,y \rangle \in C} \hat{p}_{xy} &= \sum_{\langle x,y \rangle \in C} \hat{p}_x \cdot \hat{p}_y \\
&= \sum_{x \in C_X} \hat{p}_x \sum_{\langle x,y \rangle \in C} \hat{p}_y \\
&\geq \sum_{x \in C_X} \hat{p}_x \sum_{B | \{x\} \times B \subseteq C} m_Y(B) \\
&= \sum_{x \in C_X} \sum_{B | \{x\} \times B \subseteq C} \hat{p}_x \cdot m_Y(B) \\
&= \sum_{B \subseteq A_Y} \sum_{x | \{x\} \times B \subseteq C} m_Y(B) \cdot \hat{p}_x \\
&\geq \sum_{B \subseteq A_Y} m_Y(B) \sum_{A \times B \subseteq C} m_X(A) \\
&= \sum_{A \times B \subseteq C} m_X(A) \cdot m_Y(B) \quad .
\end{aligned}
$$

This implies that

$$AU(\mathrm{Bel}) \geq - \sum_{x \in X} \sum_{y \in Y} \hat{p}_{xy} \log_2 \hat{p}_{xy}$$

$$= \ -\sum_{x \in X} \hat{p}_x \log_2 \hat{p}_x - \sum_{y \in Y} \hat{p}_y \log_2 \hat{p}_y$$

$$= \ AU(\text{Bel}_X) + AU(\text{Bel}_Y).$$

General Algorithm for Computing Function AU

Since function AU is defined in terms of the solution to a nonlinear optimization problem, its practical utility was initially questioned. Fortunately, a relatively simple and fully general algorithm for computing the measure was developed by Meyerowitz et al. [1994]. The algorithm is formulated as follows.

Algorithm 3.1 *Calculating AU from a belief function*

Input: a frame of discernment X, a belief measure Bel on X.

Output: AU (Bel), $\langle p_x \mid x \in X \rangle$ such that AU (Bel) $= - \sum_{x \in X} p_x \log_2 p_x$,
$p_i \geq 0$, $\sum_{x \in X} p_x = 1$, and Bel $(A) \leq \sum_{x \in A} p_x$ for all $\emptyset \neq A \subseteq X$.

Step 1. Find a non-empty set $A \subseteq X$, such that $\frac{\text{Bel}(A)}{|A|}$ is maximal. If there are more such sets A than one, take the one with maximal cardinality.

Step 2. For $x \in A$, put $p_x = \frac{\text{Bel}(A)}{|A|}$.

Step 3. For each $B \subseteq X - A$, put Bel $(B) = \text{Bel}(B \cup A) - \text{Bel}(A)$.

Step 4. Put $X = X - A$.

Step 5. If $X \neq \emptyset$ and Bel $(X) > 0$, then go to Step 1.

Step 6. If Bel $(X) = 0$ and $X \neq \emptyset$, then put $p_x = 0$ for all $x \in X$.

Step 7. Calculate AU (Bel) $= - \sum_{x \in X} p_x \log_2 p_x$.

The correctness of the algorithm is stated in Theorem 3.18, which was proved by Harmanec et al. [1996]. The proof employs the following two lemmas.

Lemma 3.16 *[Maeda et al., 1993] Let $x > 0$ and $c - x > 0$. Denote*

$$L(x) = - [(c - x) \log_2 (c - x) + x \log_2 x] . \tag{3.229}$$

Then $L(x)$ is strictly increasing in x when $(c - x) > x$.

Proof. $L'(x) = \log_2(c - x) - \log_2 x$ so that $L'(x) > 0$ whenever $(c - x) > x$. ∎

Lemma 3.17 *[Dempster, 1967b] Let X be a frame of discernment, Bel a generalized belief function[2] on X, and m the corresponding generalized basic probability assignment; then, a tuple $\langle p_x \mid x \in X \rangle$ satisfies the constraints*

$$0 \leq p_x \leq 1 \quad \forall\, x \in X, \tag{3.230}$$

$$\sum_{x \in X} p_x = \mathrm{Bel}(X), \tag{3.231}$$

and

$$\mathrm{Bel}(A) \leq \sum_{x \in A} p_x \quad \forall A \subseteq X. \tag{3.232}$$

if and only if there exist non-negative real numbers α_x^A for all non-empty sets $A \subseteq X$ and all $x \in A$ such that

$$p_x = \sum_{A \mid x \in A} \alpha_x^A \tag{3.233}$$

and

$$\sum_{x \mid x \in A} \alpha_x^A = m(A). \tag{3.234}$$

Using results of Lemma 3.16 and Lemma 3.17, we can now address the principal issue of this section, the correctness of Algorithm 3.1. This issue is the subject of the following theorem.

Theorem 3.18 *Algorithm 3.1 stops after finite number of steps and the output is the correct value of function AU (Bel) since $\langle p_x \mid x \in X \rangle$ maximizes the Shannon entropy within the constraints induced by Bel.*

Proof. The frame of discernment X is a finite set and the set A chosen in Step 1 of the algorithm is non-empty (note also that the set A is determined uniquely). Therefore, the "new" X has a smaller number of elements than the "old" X and so the algorithm terminates after finitely many passes through the loop of Steps 1–5.

To prove the correctness of the algorithm, we proceed in two stages. In the first stage, we show that any distribution $\langle p_x \mid x \in X \rangle$ that maximizes the Shannon entropy within the constraints induced by Bel has to satisfy for all $x \in A$ the equality $p_x = \frac{\mathrm{Bel}(A)}{|A|}$. In the second stage we show that, for any partial distribution $\langle p_x \mid x \in X - A \rangle$ that satisfies the constraints

---

[2] A generalized belief function, Bel, is a function that satisfies all requirements of a belief measure except the requirement that $\mathrm{Bel}(X) = 1$. Similarly, values of a generalized basic probability assignment are not required to add to one.

induced by the "new" generalized belief function defined in Step 3 of the algorithm, the complete distribution $\langle q_x \mid x \in X \rangle$, defined by

$$q_x = \frac{\text{Bel}(A)}{|A|}, \text{ for } x \in A, \text{ and}$$
$$q_x = p_x, \text{ for } x \in X - A, \tag{3.235}$$

satisfies the constraints induced by the original (generalized) belief function, and vice versa. That is, for any distribution $\langle q_x \mid x \in X \rangle$ satisfying constraints induced by Bel and such that $q_x = \frac{\text{Bel}(A)}{|A|}$ for all $x \in A$, it holds that $\langle q_x \mid x \in X - A \rangle$ satisfies constraints induced by the "new" generalized belief function defined in Step 3 of the algorithm. The theorem then follows by induction.

First, assume that $\langle p_x \mid x \in X \rangle$ is such that

$$AU\,(\text{Bel}) \quad = \quad -\sum_{x \in X} p_x \log_2 p_x \,, \tag{3.236}$$

$$\sum_{x \in X} p_x \quad = \quad \text{Bel}\,(X) \,, \tag{3.237}$$

and

$$\text{Bel}\,(B) \leq \sum_{x \in B} p_x \tag{3.238}$$

for all $B \subset X$, and there is $y \in A$ such that $p_y \neq \frac{\text{Bel}(A)}{|A|}$, where A is the set chosen in Step 1 of the algorithm. That is, $\frac{\text{Bel}(A)}{|A|} \geq \frac{\text{Bel}(B)}{|B|}$ for all $B \subseteq X$, and if $\frac{\text{Bel}(A)}{|A|} = \frac{\text{Bel}(B)}{|B|}$ then $|A| > |B|$. Furthermore, we may assume that $p_y > \frac{\text{Bel}(A)}{|A|}$. This is justified by the following argument: if $p_y < \frac{\text{Bel}(A)}{|A|}$, then due to $\text{Bel}\,(A) \leq \sum_{x \in A} p_x$ there exists $y' \in A$ such that $p_{y'} > \frac{\text{Bel}(A)}{|A|}$, and we may take y' instead of y.

For a finite sequence $\{x_i\}_{i=0}^m$, where $x_i \in X$ and m is a positive integer, let $\mathcal{F}$ denote the set of all focal elements associated with Bel and let

$$\Phi\,(x_i) = \cup \{C \subseteq X \mid C \in \mathcal{F}, x_{i-1} \in C, \text{ and } \alpha_{x_{i-1}}^C > 0\} \tag{3.239}$$

for $i = 1, \dots, m$, where α_x^C are the non-negative real numbers whose existence is guaranteed by Lemma 3.17 (we fix one such set of those numbers). Let

$$D = \{x \in X \mid \exists m \text{ non-negative integer and } \{x_i\}_{i=0}^m \text{ such that}$$
$$x_0 = y, \ x_m = x, \text{ for all } i = 1, 2, \dots, m, \ x_i \in \Phi\,(x_i) \text{ and for all}$$
$$i = 2, 3, \dots, m, \ \forall z \in \Phi\,(x_{i-1}) \ p_z \geq p_{x_{i-2}}\}.$$

There are two possibilities now. Either there is $z \in D$ such that in the sequence $\{x_i\}_{i=0}^m$ from the definition of D, where $x_m = z$, it holds that

$p_z < p_{m-1}$. This however leads to a contradiction with the maximality of $\langle p_x \mid x \in X \rangle$ since $- \sum_{x \in X} p_x \log_2 p_x < - \sum_{x \in X} q_x \log_2 q_x$ by Lemma 3.16, where $q_x = p_x$ for $x \in (X - \{z, x_{m-1}\})$, $q_z = p_z + \varepsilon$, and $q_{x_{m-1}} = p_{x_{m-1}} - \varepsilon$, where $\varepsilon \in \left(0, \min \left(\alpha_{x_{m-1}}^C, \frac{p_{x_{m-1}} + p_z}{2} \right) \right]$, where C is the focal element of Bel from the definition of D containing both z and x_{m-1} and such that $\alpha_{x_{m-1}}^C > 0$. The distribution $\langle q_x \mid x \in X \rangle$ satisfies the constraints induced by Bel due to the Lemma 3.17. Or, the second possibility is that for all $x \in D$ and all focal elements $C \subseteq X$ of Bel such that $x \in C$, whenever $\alpha_x^C > 0$ it holds that $p_z \geq p_x$ for all $z \in (C - \{x\})$. It follows from Lemma 3.17 that $\text{Bel}(D) = \sum_{x \in D} p_x$. However, this fact contradicts the choice of A, since

$$\frac{\text{Bel}(D)}{|D|} = \frac{\sum_{x \in D} p_x}{|D|} \geq p_y > \frac{\text{Bel}(A)}{|A|}. \tag{3.240}$$

We have shown that any $\langle p_x \mid x \in X \rangle$ maximizing the Shannon entropy within the constraints induced by Bel, has to satisfy $p_x = \frac{\text{Bel}(A)}{|A|}$ for all $x \in A$.

Let Bel' denote the generalized belief function on $X - A$ defined in Step 3 of the algorithm; i.e., $\text{Bel}'(B) = \text{Bel}(B \cup A) - \text{Bel}(A)$. It is really a generalized belief function. The reader can verify that its corresponding generalized basic probability assignment can be expressed by

$$m'(B) = \sum_{C \subseteq X \mid (C \cap (X - A)) = B} m(C) \tag{3.241}$$

for all non-empty sets $B \subseteq X - A$, and $m'(\emptyset) = 0$. Assume, then, that $\langle p_x \mid x \in X - A \rangle$ is such that $p_x \in [0, 1]$,

$$\sum_{x \in X - A} p_x = \text{Bel}'(X - A) , \tag{3.242}$$

and

$$\sum_{x \in B} p_x \geq \text{Bel}'(B) \tag{3.243}$$

for all $B \subset X - A$. Let $\langle q_x \mid x \in X \rangle$ denote the complete distribution defined by (3.235). Clearly, $q_x \in [0, 1]$. Since

$$\text{Bel}'(X - A) = \text{Bel}(X) - \text{Bel}(A) , \tag{3.244}$$

we have

$$
\begin{aligned}
\text{Bel}(X) &= \sum_{x \in X - A} p_x + \text{Bel}(A) \tag{3.245} \\
&= \sum_{x \in X - A} p_x + \sum_{x \in A} \frac{\text{Bel}(A)}{|A|} \\
&= \sum_{x \in X} q_x.
\end{aligned}
$$

From $\frac{\text{Bel}(A)}{|A|} \geq \frac{\text{Bel}(B)}{|B|}$ for all $B \subseteq X$, it follows that $\sum_{x \in B} q_x \geq \text{Bel}(C)$ for all $C \subseteq A$. Assume that $B \subseteq X$ and $B \cap (X - A) \neq \emptyset$. We know that

$$\sum_{x \in B \cap (X-A)} p_x \geq \text{Bel}'(B \cap (X - A)) = \text{Bel}(A \cup B) - \text{Bel}(A). \qquad (3.246)$$

From (2.42), we get

$$\text{Bel}(A \cup B) - \text{Bel}(A) \geq \text{Bel}(B) - \text{Bel}(A \cap B). \qquad (3.247)$$

Since $\frac{\text{Bel}(A)}{|A|} \geq \frac{\text{Bel}(A \cap B)}{|A \cap B|}$, we get

$$\sum_{x \in B} q_x = \sum_{x \in B \cap (X-A)} p_x + \sum_{x \in A \cap B} \frac{\text{Bel}(A)}{|A|} \geq \text{Bel}(B). \qquad (3.248)$$

Conversely, assume $\langle q_x \mid x \in X \rangle$ is such that $q_x \in [0, 1]$, $q_x = \frac{\text{Bel}(A)}{|A|}$ for all $x \in A$, $\sum_{x \in X} q_x = \text{Bel}(X)$, and $\sum_{x \in B} q_x \geq \text{Bel}(B)$ for all $B \subset X$. Clearly, we have

$$\text{Bel}'(X - A) = \text{Bel}(X) - \text{Bel}(A) = \sum_{x \in X - A} q_x. \qquad (3.249)$$

Let $C \subseteq X - A$. We know that $\sum_{x \in A \cup C} q_x \geq \text{Bel}(A \cup C)$, but it follows from this fact that

$$\sum_{x \in C} q_x \geq \text{Bel}(A \cup C) - \text{Bel}(A) = \text{Bel}'(C). \qquad (3.250)$$

This means that we reduced the size of our problem and, therefore, the theorem follows by induction. ∎

Several remarks regarding Algorithm 3.1 can be made:

(a) Note that we have also proved that the distribution maximizing the entropy within the constraints induced by a given belief function is unique. It is possible to prove this fact directly by using the concavity of the Shannon entropy.

(b) The condition that A has the maximal cardinality among sets B with equal value of $\frac{\text{Bel}(B)}{|B|}$ in Step 1 is not necessary for the correctness of the algorithm, but it speeds it up. The same is true for the condition $\text{Bel}(X) > 0$ in Step 5. Moreover, we could exclude the elements of X outside the union of all focal elements of Bel (usually called *core* within evidence theory) altogether.

(c) If $A \subset B$ and $\text{Bel}(A) = \text{Bel}(B)$, then $\frac{\text{Bel}(A)}{|A|} > \frac{\text{Bel}(B)}{|B|}$. It means that it is not necessary to work with the whole power set $\mathcal{P}(X)$; it is enough to consider $\mathcal{C} = \{A \subseteq X \mid \exists \{F_1, \ldots, F_l\} \subseteq \mathcal{P}(X)$ such that $m(F_i) > 0$ and $A = \cup_{i=1}^{l} F_i\}$.

TABLE 3.2 The values of $Bel(A)$ and $\frac{Bel(A)}{|A|}$ in Example 3.1.

A	$\mathrm{Bel}(A)$	$\frac{\mathrm{Bel}(A)}{|A|}$
$\{a\}$	0.26	0.26
$\{b\}$	0.26	0.26
$\{c\}$	0.26	0.26
$\{a,b\}$	0.59	0.295
$\{a,c\}$	0.53	0.265
$\{a,d\}$	0.27	0.135
$\{b,c\}$	0.53	0.265
$\{b,d\}$	0.27	0.135
$\{c,d\}$	0.27	0.135
$\{a,b,c\}$	0.87	0.29
$\{a,b,d\}$	0.61	$0.20\overline{3}$
$\{a,c,d\}$	0.55	$0.18\overline{3}$
$\{b,c,d\}$	0.55	$0.18\overline{3}$
$\{a,b,c,d\}$	1	0.25

Example 3.1 *Given the frame of discernment $X = \{a,b,c,d\}$, let a belief function* Bel *be defined by the associated basic probability assignment m :*

$$m(\{a\}) = 0.26,$$
$$m(\{b\}) = 0.26,$$
$$m(\{c\}) = 0.26,$$
$$m(\{a,b\}) = 0.07,$$
$$m(\{a,c\}) = 0.01,$$
$$m(\{a,d\}) = 0.01,$$
$$m(\{b,c\}) = 0.01,$$
$$m(\{b,d\}) = 0.01,$$
$$m(\{c,d\}) = 0.01,$$
$$m(\{a,b,c,d\}) = 0.1$$

(only values for focal elements are listed). By Remark (c), we do not need to consider the value of Bel *for $\{d\}$ and $\emptyset$. The values of* Bel(A) *and also the values of $\frac{Bel(A)}{|A|}$ for all other subsets A of X are listed in Table 3.2. We can see from this table that the value $\frac{Bel(\{a,b\})}{|\{a,b\}|}$ is highest for $\{a,b\}$. Therefore, we set $p_a = p_b = 0.295$. Now, we have to update our (now generalized) belief function* Bel*. For example,*

$$\begin{aligned} \mathrm{Bel}(\{c\}) &= \mathrm{Bel}(\{a,b,c\}) - \mathrm{Bel}(\{a,b\}) \\ &= 0.87 - 0.59 \end{aligned}$$

TABLE 3.3 Values of Bel and $\frac{Bel(A)}{|A|}$ in the second pass through the loop in Example 3.1.

A	$Bel(A)$	$\frac{Bel(A)}{|A|}$
$\{c\}$	0.28	0.28
$\{d\}$	0.02	0.02
$\{c, d\}$	0.41	0.205

$$= \quad 0.28.$$

All the values are listed in Table 3.3. Our new frame of discernment X is now $\{c, d\}$. Since $X \neq \emptyset$ and $Bel(X) > 0$ we repeat the process. The maximum of $\frac{Bel(A)}{|A|}$ is now reached at $\{c\}$. We put $p_c = 0.28$, change $Bel(\{d\})$ to 0.13, and X to $\{d\}$. In the last pass through the loop we get $p_d = 0.13$. We may conclude that

$$
\begin{aligned}
AU\,(Bel) \;&=\; -\sum_{i \in \{a,b,c,d\}} p_i \log_2 p_i \\
&=\; -2 \times 0.295 \log_2 0.295 - 0.28 \log_2 0.28 - 0.13 \log_2 0.13 \\
&\doteq\; 1.93598.
\end{aligned}
$$

Computing Function AU in Possibility Theory

Due to the nested structure of possibilistic bodies of evidence, the computation of function AU can be substantially simplified. It is thus useful to reformulate Algorithm 3.1 in terms of possibility distribution for applications in possibility theory. The following is the reformulated algorithm.

Algorithm 3.2 *Calculating AU from a possibility distribution*

Input: $n \in \mathbb{N}$, $\mathbf{r} = \langle r_1, r_2, \ldots, r_n \rangle$.

Output: $AU\,(Pos)$, $\langle p_i \mid i \in \mathbb{N}_n \rangle$ such that $AU\,(Pos) = -\sum_{i=1}^{n} p_i \log_2 p_i$, with $p_i \geq 0$ for $i \in \mathbb{N}_n$, $\sum_{i=1}^{n} p_i = 1$, and $\sum_{x \in A} p_x \leq Pos\,(A)$ for all $\emptyset \neq A \subset X$.

Step 1. Let $j = 1$ and $r_{n+1} = 0$.

Step 2. Find maximal $i \in \{j, j+1, \ldots, n\}$ such that $\frac{r_j - r_{i+1}}{i+1-j}$ is maximal.

Step 3. For $k \in \{j, j+1, \ldots, i\}$ put $p_k = \frac{r_j - r_{i+1}}{i+1-j}$.

Step 4. Put $j = i + 1$.

Step 5. If $i < n$, then go to Step 2.

Step 6. Calculate $AU(\text{Pos}) = -\sum_{i=1}^{n} p_i \log_2 p_i$.

As already mentioned, it is sufficient to consider only all unions of focal elements of a given belief function, in this case a necessity measure. Since all focal elements of a necessity measure are nested, the union of a set of focal elements is the largest focal element (in the sense of inclusion) in the set. Therefore, we have to examine the values of $\frac{\text{Nec}(A)}{|A|}$ only for A being a focal element.

We show by induction on number of passes through the loop of Steps 1 – 5 of Algorithm 3.1 that the following properties hold in a given pass:

(a) the "current" frame of discernment X is the set $\{x_j, x_{j+1}, \ldots, x_n\}$, where the value of j is taken in the corresponding pass through the loop of Steps 2 – 5 of Algorithm 3.2,

(b) all focal elements are of the form $A_i = \{x_j, x_{j+1}, \ldots, x_i\}$ for some $i \in \{j, j+1, \ldots, n\}$, where j has the same meaning as in (a), and

(c) $\text{Nec}(A_i) = r_j - r_{i+1}$, where j is again as described in (a).

This implies that Algorithm 3.2 is a correct modification of Algorithm 3.1 for the case of possibility theory.

In the first pass, $j = 1$, $X = \{x_1, x_2, \ldots, x_n\}$; (b) holds due to our ordering convention regarding possibility distribution. Since $r_1 = 1$, we have

$$
\begin{aligned}
\text{Nec}(A_i) &= 1 - \text{Pos}\left(\overline{A_i}\right) & (3.251) \\
&= 1 - \max_{k=i+1}^{n} r_j \\
&= 1 - r_{i+1}.
\end{aligned}
$$

So (c) is true.

Let us assume now that (a) – (c) were true in some fixed pass. We want to show that (a) – (c) hold in the next pass. Let l denote the value of i maximizing $\frac{\text{Nec}(A_i)}{|A_i|} = \frac{\text{Nec}(A_j)}{i+1-j}$. X now becomes $X - A_l = \{x_{l+1}, x_{l+2}, \ldots, x_n\}$. Therefore (a) holds, since $j = l+1$ and $\text{Nec}(A_i)$ becomes

$$
\begin{aligned}
\text{Nec}(A_i) &= \text{Nec}(A_l \cup A_i) - \text{Nec}(A_l) & (3.252) \\
&= \left[1 - \text{Pos}\left(\overline{A_l \cup A_i}\right)\right] - \left[1 - \text{Pos}\left(\overline{A_l}\right)\right] \\
&= \left[1 - \max_{k=i+1}^{n} r_j\right] - \left[1 - \max_{k=l+1}^{n} r_j\right] \\
&= r_{l+1} - r_{i+1} \\
&= r_j - r_{j+1}
\end{aligned}
$$

for $i \in \{j, j+1, \ldots, n\}$. This implies that (b) and (c) hold.

TABLE 3.4 The values of $\frac{r_j - r_{i+1}}{i+1-j}$ in Example 3.2.

Pass	$\frac{r_j - r_{i+1}}{i+1-j}$	
	1	2
$i \setminus j$	1	4
1	0.1	
2	0.075	
3	0.1$\overline{6}$	
4	0.125	0
5	0.13	0.075
6	0.11$\overline{6}$	0.0$\overline{6}$
7	0.1286	0.1
8	0.125	0.1

Example 3.2 *Consider* $X = \{1, 2, \ldots, 8\}$ *and the possibility distribution*

$$\mathbf{p} = \langle 1, 0.9, 0.85, 0.5, 0.5, 0.35, 0.3, 0.1 \rangle \, .$$

The relevant values of $\frac{r_j - r_{i+1}}{i+1-j}$ *are listed in Table 3.4.*

We can see there that in the first pass the maximum is reached at $i = 3$, and $p_1 = p_2 = p_3 = \frac{1}{6}$. In second pass the maximum is reached at both $i = 7$ and $i = 8$. We take the bigger one and put $p_4 = p_5 = p_6 = p_7 = p_8 = 0.1$. We finish by computing

$$
\begin{aligned}
AU \, (\text{Pos}) &= -\sum_{i=1}^{8} p_i \log_2 p_i = \frac{3}{6} \log_2 6 + \frac{5}{10} \log_2 10 \\
&= 2.953445298
\end{aligned}
$$

3.4 Fuzziness

The third type of uncertainty is fuzziness (or vagueness). In general, a measure of fuzziness is a function

$$f : \tilde{\mathcal{P}}(X) \to \mathbb{R}^+ \tag{3.253}$$

where $\tilde{\mathcal{P}}(X)$ denotes the set of all fuzzy subsets of X (fuzzy power set). For each fuzzy set A, this function assigns a nonnegative real number $f(A)$ that expresses the degree to which the boundary of A is not sharp.

In order to qualify as a sensible measure of fuzziness, function f must satisfy some requirements that adequately capture our intuitive comprehension of the degree of fuzziness. The following three requirements are essential:

(f1) Boundary condition — $f(A) = 0$ iff A is a crisp set.

(f2) Maximum — $f(A)$ attains its maximum iff A is maximally fuzzy.

One way to define the fuzzy set that maximizes fuzziness is to require that $A(x) = 0.5$ for all $x \in X$, which is intuitively conceived as the highest fuzziness.

(f3) Sharpness Monotonicity — $f(A) \leq f(B)$ when set A is undoubtedly sharper than set B, $A \prec B$.

According to our intuition, sharpness means that

$$A(x) \leq B(x) \text{ whenever } B(x) \leq 0.5 \qquad (3.254)$$

and

$$A(x) \geq B(x) \text{ whenever } B(x) \geq 0.5 \qquad (3.255)$$

for all $x \in X$.

There are different ways of measuring fuzziness that all satisfy the three essential requirements. One way is to measure fuzziness of any set A by a metric distance between its membership grade function, A, and the membership grade function (or characteristic function) of the nearest crisp set [Kaufmann, 1975]. Even when committing to this conception of measuring fuzziness, the measurement is not unique. To make it unique, we have to choose one of many possible distance functions.

Another way of measuring fuzziness, which seems more practical as well as more general, is to view fuzziness of a set in terms of the lack of distinction between the set and its complement [Yager, 1979, 1980b]. Indeed, it is precisely the lack of distinction between sets and their complements that distinguishes fuzzy sets from crisp sets. The less a set differs from its complement, the fuzzier it is. In this section, we restrict ourselves to this view of fuzziness, which is currently predominant in the literature.

Measuring fuzziness in terms of distinctions between sets and their complements is dependent on the definition of a fuzzy complement [Klir and Folger, 1988]. This dependence was thoroughly investigated by Higashi and Klir [1982].

Employing the standard fuzzy complement, we can still choose different distance functions to express the lack of distinction of a set and its complement. Choosing, for example, the Hamming distance, the local distinction (one for each $x \in X$) of a given set A and its standard complement, $\bar{A}$, is measured by

$$\left| A(x) - \bar{A}(x) \right| = \left| 2A(x) - 1 \right|, \qquad (3.256)$$

and the lack of each local distinction is measured by

$$1 - \left| 2A(x) - 1 \right|. \qquad (3.257)$$

FIGURE 3.4 A fuzzy set and a crisp set with equal nonspecificity.

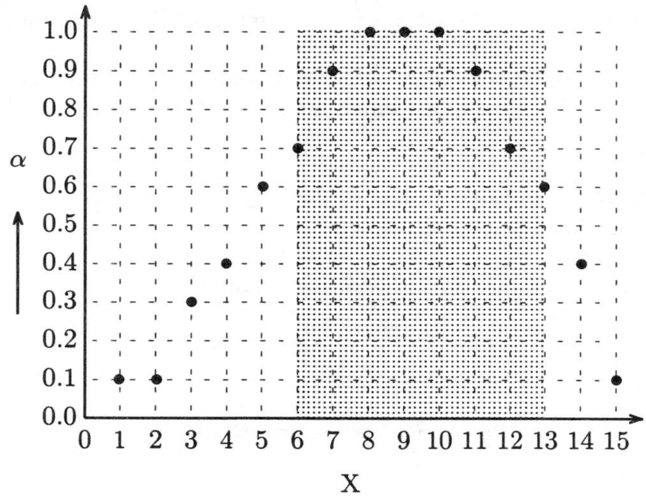

The measure of fuzziness, $f(A)$, is then obtained by adding all these local measurements:

$$f(A) = |X| - \sum_{x \in X} |2A(x) - 1|. \qquad (3.258)$$

Since $f(A) = 1$ when $A(x) = 0.5$ for one particular element $x \in X$ and $A(x) \in \{0,1\}$ for $y \neq x$, the degree of fuzziness measured by f is expressed in bits. Here, one bit of fuzziness is equivalent to the total perplexity whether one particular element of X belongs to a set or to its complement. The range of function f is $[0, |X|]$; $f(A) = 0$ if and only if A is a crisp set, $f(A) = X$ when $A(x) = 0.5$ for all $x \in X$.

If we allow for an arbitrary complement operator, co, then we must alter our definitions of maximally fuzzy and of sharpness in the formulation of the Axioms (f2) and (f3). If the fuzzy complement operator co has an equilibrium point, e_{co}, where $co(e_{co}) = e_{co}$, then Axiom (f2) should be altered so that a fuzzy set A with $A(x) = e_{co}$ for all $x \in X$ has the greatest fuzziness value of any fuzzy set on X. Similarly Axiom (f3) must be altered so that,

$$A \prec B \quad \text{if and only if} \quad |A(x) - co(A(x))| \geq |B(x) - co(B(x))| \quad (3.259)$$

for all $x \in X$. If we aggregate the distance,

$$\delta[co, A](x) = |A(x) - co(A(x))|, \qquad (3.260)$$

between an element's membership in A and its membership in the complement of A over all the elements of X using one of the Minkowski class of

FIGURE 3.5 Three basic types of uncertainty.

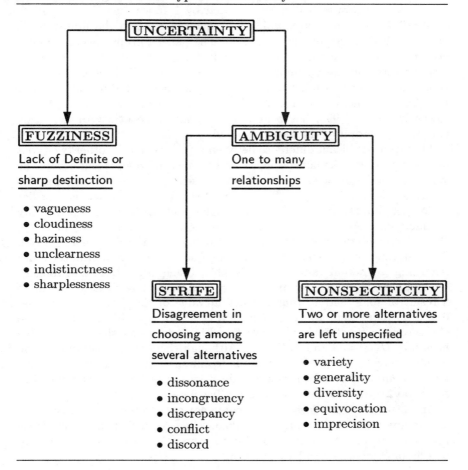

metrics with parameter $w \in [1, \infty]$ we arrive at the formulation,

$$D[w, co](A) = \left(\sum_{x \in X} (\delta_{co,A}(x))^w \right)^{1/w} . \qquad (3.261)$$

The generalized measure of fuzziness then has the form.

$$f[D](A) = D(Z) - D(A) \qquad (3.262)$$

where Z is an arbitrary crisp subset of X that maximizes the distance $D[w, co]$.

An alternative measure of fuzziness,

$$^{e}f(A) = - \sum_{x \in X} \left[A(x) \log_2 A(x) + \bar{A}(x) \log_2 \bar{A}(x) \right] \qquad (3.263)$$

based on a different distance function within the same general view of fuzziness, was proposed and investigated by DeLuca and Termini [1972, 1974, 1977]. This measure, which is frequently cited in the literature, is usually called an entropy of fuzzy sets. The name entropy was apparently chosen due to the similarity of the product terms in Eq. (3.263) with the product terms in the Shannon entropy (Eq. (3.107)). This choice is unfortunate since the two functions measure fundamentally different types of uncertainty. Function $^e f$ measures fuzziness in bits and its range is $[0, |X|]$. Recently Kosko [1993] has begun to investigate the utility of the fuzzy entropy in relation to the measure of subsethood.

There is a significant difference between fuzziness and the other two types of uncertainty. Unlike nonspecificity and conflict, fuzziness results from the vagueness of linguistic expressions. In some sense, however, fuzziness is also connected with information. To illustrate this connection, let us consider an integer-valued variable whose exact value is of our interest, but the only available information regarding its value is given in terms of the fuzzy set A whose membership grade function is defined by the dots in Fig. (3.4). This information is incomplete and, consequently, it results in uncertainty regarding the actual value of the variable. Using Eqs. (3.31) and (3.258), we find that $U(A) = 3.05$ and $f(A) = 6$. Assume now that we learn, as a result of some action, that the actual value of the variable is at least 6 and no more than 13. This defines a crisp set $B = \{6, 7, ..., 13\}$ whose characteristic function is illustrated in Fig. 3.4 by the shaded area. We have now $U(B) = 3$ and $f(B) = 0$. Although there is virtually no change in nonspecificity, the action helped us to eliminate fuzziness. We may thus view the amount of information obtained by the action as proportional in some way to the amount of eliminated fuzziness.

Any measure of fuzziness is applicable not only to fuzzy set theory, but also to any other theory that is fuzzified. When focal elements in evidence theory are fuzzy sets, it is reasonable to express the degree of fuzziness of a body of evidence $\langle \mathcal{F}, m \rangle$ by function

$$\tilde{f}(m) = \sum_{A \in \mathcal{F}} m(A) \, f(A) \qquad (3.264)$$

which represents the weighted average of the degree of fuzziness of focal elements.

3.5   Summary of Uncertainty Measures

The three types of uncertainty whose measures are now well established in classical set theory, fuzzy set theory, probability theory, possibility theory, and evidence theory are summarized in Fig. 3.5. Each type is depicted by a brief common-sense characterization and a group of pertinent synonyms.

Two of the uncertainty types, *nonspecificity* and *strife,* are viewed as species of a higher uncertainty type, which seems well captured by the term *ambiguity,* the latter is associated with any situation in which it remains unclear which of several alternatives should be accepted as the genuine one. In general, ambiguity results from the lack of certain distinctions characterizing an object (nonspecificity), from conflicting distinctions, or from both of these. The third uncertainty type, *fuzziness,* is different from ambiguity; it results from the lack of sharpness of relevant distinctions.

Table 3.5 contains functions by which the various types of uncertainty are measured. For each function, four pieces of information are given in the table: the type of uncertainty measured by the function, the uncertainty theory to which the function applies, the year in which the function was proposed as a measure of the respective uncertainty, and the number of equation by which the function is defined. Measures in Table 3.5 are applicable only to finite sets. Various uncertainty measures for infinite sets are given by Eqs. (3.78), (3.103 – 3.105), and (3.184).

Uncertainty theory	Uncertainty measure	Eq. Year	Uncertainty type Name
Classical set theory	$H(A) = log_2|A|$	(3.8) 1928	Nonspecificity Hartley function
Fuzzy set theory	$U(A) = \int\limits_0^{h(A)} \log_2 |^\alpha A|\, d\alpha + (1 - h(A)) \log_2 |X|$	(3.38) 1995	Nonspecificity U-uncertainty
Possibility theory	$U(r) = \sum\limits_{i=2}^{n} r_i \log_2 \frac{i}{i-1} + (1 - r_1) \log_2 |n|$	(3.39) 1995	Nonspecificity U-uncertainty
Evidence theory	$N(m) = \sum\limits_{A \in \mathcal{F}}^{n} m(A) \log_2 |A|$	(3.51) 1985	Nonspecificity Generalized measure
Probability theory	$S(p) = - \sum\limits_{x \in X} p(x) \log_2 p(x)$	(3.107) 1948	Conflict Shannon entropy
Evidence theory	$E(m) = - \sum\limits_{A \in \mathcal{F}} m(A) \log_2 Pl(A)$	(3.187) 1983	Conflict Dissonance
Evidence theory	$C(m) = - \sum\limits_{A \in \mathcal{F}} m(A) \log_2 Bel(A)$	(3.188) 1982	Conflict Confusion
Evidence theory	$D(m) = - \sum\limits_{A \in \mathcal{F}} m(A) \log_2 \sum\limits_{B \in \mathcal{F}} m(B) \frac{|A \cap B|}{|B|}$	(3.202) 1977	Conflict Discord
Evidence theory	$ST(m) = - \sum\limits_{A \in \mathcal{F}} m(A) \log_2 \sum\limits_{B \in \mathcal{F}} m(B) \frac{|A \cap B|}{|A|}$	(3.206) 1992	Conflict Strife
Possibility theory	$ST(r) = \sum\limits_{i=2}^{n} (r_i - r_{i+1}) \log_2 \frac{i}{\sum_{j=1}^{i} r_j}$	(3.211) 1992	Conflict Strife
Evidence theory	$AU(Bel) = \max\limits_{\mathcal{P}_{Bel}} \left[- \sum\limits_{x \in X} p(x) \log_2 p(x) \right]$	(3.217) 1995	Total Aggregate uncertainty
Fuzzy set theory	$^e f(A) = - \sum\limits_{x \in X} [A(x) \log_2 A(x) + \bar{A}(x) \log_2 \bar{A}(x)]$	(3.263) 1972	Fuzziness Fuzzy entropy
Fuzzy set theory	$f(A) = |X| - \sum\limits_{x \in X} |2A(x) - 1|$	(3.258) 1979	Fuzziness Standard measure
Fuzzified evidence theory	$\tilde{f}(m) = \sum\limits_{A \in \mathcal{F}} m(A) f(A)$	(3.264) 1988	Fuzziness Generalized measure

TABLE 3.5: SUMMARY OF UNCERTAINTY MEASURES

4
PRINCIPLES OF UNCERTAINTY

Although measures of the various types of uncertainty-based information (Table 3.5) are not sufficient in human communication [Cherry, 1957], they are highly effective tools for dealing with systems problems of virtually any kind [Klir, 1985]. For the classical information measures (Hartley function and Shannon entropy), which were originally conceived solely as tools for analyzing and designing telecommunication systems, this broad utility is best demonstrated by Ashby [1958, 1965, 1969, 1972] and Conant [1969, 1974, 1976, 1981].

The use of uncertainty measures for dealing with systems problems is well captured by three general principles of uncertainty [Klir, 1995b]. These principles, which are the subject of this section, are: a principle of minimum uncertainty, a principle of maximum uncertainty, and a principle of uncertainty invariance. Due to the connection between uncertainty and uncertainty-based information, these principles may also be interpreted as principles of information.

4.1 Principle of Minimum Uncertainty

The *principle of minimum uncertainty* is basically an arbitration principle. It is used, in general, for narrowing down solutions in various systems problems that involve uncertainty. The principle states that we should accept only those solutions, from among all otherwise equivalent solutions, whose uncertainty (pertaining to the purpose concerned) is minimal.

A major class of problems for which the principle of minimum uncertainty is applicable are *simplification problems*. When a system is simplified, it is usually unavoidable to lose some information contained in the system. The amount of information that is lost in this process results in the increase of an equal amount of relevant uncertainty. Examples of relevant uncertainties are predictive, retrodictive, or prescriptive uncertainty. A sound simplification of a given system should minimize the loss of relevant information (or the increase in relevant uncertainty) while achieving the required reduction of complexity. That is, we should accept only such simplifications of a given system at any desirable level of complexity for which the loss of relevant information (or the increase in relevant uncertainty) is minimal. When properly applied, the principle of minimum uncertainty guarantees that no information is wasted in the process of simplification.

There are many simplification strategies, which can perhaps be classified into three main classes:

- simplifications made by eliminating some entities from the system (variables, subsystems, etc.);

- simplifications made by aggregating some entities of the system (variables, states, etc.);

- simplifications made by breaking overall systems into appropriate subsystems.

Regardless of the strategy employed, the principle of minimum uncertainty is utilized in the same way. It is an arbiter which decides which simplifications to choose at any given level of complexity or, alternatively, which simplifications to choose at any given level of acceptable uncertainty. Let us describe this important role of the principle of minimum uncertainty in simplification problems more formally.

Let Z denote a system of some type and let Q_Z denote the set of all simplifications of Z that are considered *admissible* in a given context. For example, Q_Z may be the set of simplifications of Z that are obtained by a particular simplification method. Let $\leq_c$ and $\leq_u$ denote *preference orderings* on Q_Z that are based upon complexity and uncertainty, respectively. In general, systems with smaller complexity and smaller uncertainty are preferred. The two preference orderings can be combined in a *joint preference ordering*, $\leq$, defined as follows: for any pair of systems $x, y \in Q_Z$, $x \leq y$ if and only if $x \leq_c y$ and $x \leq_u y$. The joint preference ordering is usually a partial ordering, even though it may be in some simplification problems only a weak ordering (reflexive and transitive relation on Q_Z). The use of the uncertainty preference ordering, which exemplifies in this case the principle of minimum uncertainty, enables us to narrow down all admissible simplifications to a small set of *preferred simplifications*. The latter forms a *solution set*, S_Z, of the simplification problem, which consists of those admissible simplifications in Q_Z that are either equivalent or incomparable in terms of the joint preference ordering. Formally,

$$S_Z = \{x \in Q_Z \mid \text{ for all } y \in Q_Z, \ y \leq x \text{ implies } x \leq y\}. \tag{4.1}$$

Observe that the solution set S_Z in this formulation, which may be called an *unconstrained simplification problem*, contains simplifications at various levels of complexity and with various degrees of uncertainty. The problem may be constrained, for example, by considering simplifications admissible only when their complexities are at some designated level, when they are below a specified level, when their uncertainties do not exceed certain maximum acceptable level, etc. The formulation of these various *constrained simplification problems* differs from the above formulation only in the definition of the set of admissible simplifications Q_Z.

Another application of the principle of minimum uncertainty is the area of *conflict-resolution problems*. For example, when we integrate several overlapping models into one larger model, the models may be locally inconsistent. It is reasonable then to require that each of the models be appropriately adjusted in such a way that the overall model becomes consistent. It is obvious that some information contained in the given models is inevitably lost by these adjustments. This is not desirable. Hence, we should minimize this loss of information. That is, we should accept only those adjustments for which the total loss of information (or total increase of uncertainty) is minimal. The total loss of information may be expressed, for example, by the sum of all individual losses or by a weighted sum, if the given models are valued differently.

Thus far, the principle of minimum uncertainty has been employed predominantly within the domain of probability theory, where the function to be minimized is the Shannon entropy (usually some of its conditional forms) or another function based on it (information transmission, directed divergence, etc.). The great utility of this principle in dealing with a broad spectrum of problems is perhaps best demonstrated by the work of Christensen [1980-81, 1985, 1986]. Another important user of the principle of minimum entropy is Watanabe [1981, 1985], who has repeatedly argued that entropy minimization is a fundamental methodological tool in the problem area of pattern recognition. Outside probability theory, the principle of minimum uncertainty has been explored in reconstructability analysis of possibilistic systems [Cavallo and Klir, 1982; Klir, 1985, 1990b; Klir and Parviz, 1986, Klir and Way, 1985; Mariano, 1987]; the function that is minimized in these explorations is the U-uncertainty or an appropriate function based upon it [Higashi and Klir, 1983b].

4.2   Principle of Maximum Uncertainty

The second principle, the *principle of maximum uncertainty*, is essential for any problem that involves *ampliative reasoning*. This is reasoning in which conclusions are not entailed in the given premises. Using common sense, the principle may be expressed by the following requirement: in any ampliative inference, use all information available but make sure that no additional information is unwittingly added. That is, employing the connection between information and uncertainty, the principle requires that conclusions resulting from any ampliative inference maximize the relevant uncertainty within the constraints representing the premises. This principle guarantees that our ignorance be fully recognized when we try to enlarge our claims beyond the given premises and, at the same time, that all information contained in the premises be fully utilized. In other words, the principle guarantees that our conclusions are maximally noncommittal with regard to information not contained in the premises.

Ampliative reasoning is indispensable to science in a variety of ways. For example, whenever we utilize a scientific model for predictions, we employ ampliative reasoning. Similarly, when we want to estimate microstates from the knowledge of relevant macrostates and partial information regarding the microstates (as in image processing and many other problems), we must resort to ampliative reasoning. The problem of the identification of an overall system from some of its subsystems is another example that involves ampliative reasoning.

Ampliative reasoning is also common and important in our daily life, where, unfortunately, the principle of maximum uncertainty is not always adhered to. Its violation leads almost invariably to conflicts in human communication, as well expressed by Bertrand Russell in his *Unpopular Esseys* [London, 1950]:

> ... whenever you find yourself getting angry about a difference in opinion, be on your guard; you will probably find, on examination, that your belief is getting beyond what the evidence warrants.

The principle of maximum uncertainty is well developed and broadly utilized within classical information theory, where it is called the *principle of maximum entropy*. It was founded, presumably, by Jaynes [1983] and now has a history of several decades [Kapur, 1983]. Perhaps the greatest skill in using this principle in a broad spectrum of applications has been demonstrated by Christensen [1980-81, 1985, 1986], Jaynes [1982, 1983, 1989], Kapur [1989, 1994], and Tribus [1969].

Literature concerned with the maximum entropy principle is extensive. For example, 640 references regarding the principle are listed in the above mentioned book by Kapur [1989], which is an excellent and up-to date overview of the astonishing range of utility of the principle. Let us mention just a few books of special significance: Batten [1983], Buck and Macaulay [1991], Erickson and Smith [1988], Grandy and Schick [1991], Justice [1986], Kapur and Kesavan [1987], Levine and Tribus [1979], Skilling [1989], Smith et al. [1985, 1987, 1992], Theil [1967], Theil and Fiebig [1984], Webber [1979], Wilson [1970].

A general formulation of the principle of maximum entropy is: determine a probability distribution $\langle p(x) \mid x \in X \rangle$ that maximizes the Shannon entropy subject to given constraints $c_1, c_2,...,c_n$ which express partial information about the unknown probability distribution, as well as general constraints (axioms) of probability theory. The most typical constraints employed in practical applications of the maximum entropy principle are mean (expected) values of one or more random variables or various marginal probability distributions of an unknown joint distribution.

As an example, consider a random variable x with possible (given) non-negative real values $x_1, x_2,...,x_n$. Assume that probabilities p_i of values x_i ($i \in \mathbb{N}_n$) are not known, but we know the mean (expected) value $E(x)$ of the variable. Employing the maximum entropy principle, we estimate the unknown probabilities p_i, $i \in \mathbb{N}_n$, by solving the following optimization problem:

Maximize the function

$$S(p_1, p_2, \ldots, p_n) = -\sum_{i=1}^{n} p_i \ln p_i \qquad (4.2)$$

subject to the constraints

$$E(x) = \sum_{i=1}^{n} p_i x_i \qquad (4.3)$$

and

$$p_i \geq 0 \ (i \in \mathbb{N}_n), \ \sum_{i=1}^{n} p_i = 1. \qquad (4.4)$$

For the sake of simplicity we use the natural logarithm in the definition of Shannon entropy in this formulation. Equation (4.3) represents the available information; (4.4) represents the standard constraints imposed upon p by probability theory.

First, we form the Lagrange function

$$L = -\sum_{i=1}^{n} p_i \ln p_i - \alpha \left(\sum_{i=1}^{n} p_i - 1 \right) - \beta \left(\sum_{i=1}^{n} p_i x_i - E(x) \right) \qquad (4.5)$$

where α and β are the Lagrange multipliers that correspond to the two constraints. Second, we form the partial derivatives of L with respect to p_i

$(i \in \mathbb{N}_n)$, α, and β and set them equal to zero; this results in the equations

$$\frac{\partial L}{\partial p_i} = -\ln p_i - 1 - \alpha - \beta x_i = 0 \text{ for each } i \in \mathbb{N}_n \qquad (4.6a)$$

$$\frac{\partial L}{\partial \alpha} = 1 - \sum_{i=1}^{n} p_i = 0 \qquad (4.6b)$$

$$\frac{\partial L}{\partial \beta} = E(x) - \sum_{i=1}^{n} p_i \, x_i = 0 . \qquad (4.6c)$$

The last two equations are exactly the same as the constraints of the optimization problem. The first n equations can be written as

$$\begin{aligned} p_1 &= e^{-1-\alpha-\beta x_1} = e^{-(1+\alpha)} e^{-\beta x_1} \\ p_2 &= e^{-1-\alpha-\beta x_2} = e^{-(1+\alpha)} e^{-\beta x_2} \\ &\vdots \qquad \vdots \\ p_n &= e^{-1-\alpha-\beta x_n} = e^{-(1+\alpha)} e^{-\beta x_n} \end{aligned} \qquad (4.7)$$

When we divide each of these equations by the sum of all of them (which must be one), we obtain

$$p_i = \frac{e^{-\beta x_i}}{\sum_{k=1}^{n} e^{-\beta x_k}} \qquad (4.8)$$

for each $i \in \mathbb{N}_n$. In order to determine the value of β, we multiply the ith equation of (4.8) by x_i and add all of the resulting equations, thus obtaining

$$E(x) = \frac{\sum_{i=1}^{n} x_i \, e^{-\beta x_i}}{\sum_{i=1}^{n} e^{-\beta x_i}} \qquad (4.9)$$

and

$$\sum_{i=1}^{n} x_i \, e^{-\beta x_i} - E(x) \sum_{i=1}^{n} e^{-\beta x_i} = 0 . \qquad (4.10)$$

Multiplying this equation by $e^{\beta \, E(x)}$ results in

$$\sum_{i=1}^{n} [x_i - E(x)] \, e^{-\beta [x_i - E(x)]} = 0 . \qquad (4.11)$$

This equation must now be solved (numerically) for β and the solution substituted for β in Eq. (4.8), which results in the estimated probabilities p_i $(i \in \mathbb{N}_n)$.

Example 4.1 *Consider first an "honest" (unbiased) die. Here $x_i = i$ for $i \in \mathbb{N}_6$ and $E(x) = 3.5$. Equation (4.11) has the form*

$$-2.5e^{2.5\beta} - 1.5e^{1.5\beta} - 0.5e^{0.5\beta} + 0.5e^{-0.5\beta} + 1.5e^{-1.5\beta} + 2.5e^{-2.5\beta} = 0.$$

The solution is clearly $\beta = 0$; when this value is substituted for β in Eq. (4.8), we obtain the uniform probability $p_i = \frac{1}{6}$ for all $i \in N_6$.

Consider now a biased die for which it is known that $E(x) = 4.5$. Equation (4.11) assumes a different form

$$-3.5e^{3.5\beta} - 2.5e^{2.5\beta} - 1.5e^{1.5\beta} - 0.5e^{0.5\beta} + 0.5e^{-0.5\beta} + 1.5e^{-1.5\beta} = 0.$$

When solving this equation (by a suitable numerical method), we obtain $\beta = -0.37105$. Substitution of this value for β into Eq. (4.8) yields the maximum entropy probability distribution

$$p_1 = \frac{1.45}{26.66} = 0.05$$

$$p_2 = \frac{2.10}{26.66} = 0.08$$

$$p_3 = \frac{3.04}{26.66} = 0.11$$

$$p_4 = \frac{4.41}{26.66} = 0.17$$

$$p_5 = \frac{6.39}{26.66} = 0.24$$

$$p_6 = \frac{9.27}{26.66} = 0.35$$

Our only knowledge about the random variable x in this example is the knowledge of its expected value $E(x)$. It is expressed by Eq. (4.3) as a constraint on the set of relevant probability distributions. If $E(x)$ were not known, we would be totally ignorant about x and the maximum entropy principle would yield the uniform probability distribution (the only distribution for which the entropy reaches its absolute maximum). The entropy of the probability distribution given by Eq. (4.8) is smaller than the entropy of the uniform distribution, but it is the largest entropy from among all the entropies of the probability distributions that conform to the given expected value $E(x)$.

A generalization of the principle of maximum entropy is the *principle of minimum cross-entropy* [Williams, 1980; Shore and Johnson, 1981]. It can be formulated as follows: given a prior probability distribution function q on a finite set X and some relevant new evidence, determine a new probability distribution function p that minimizes the cross-entropy $\hat{S}$ given by Eq. (3.185) subject to constraints $c_1, c_2,...,c_n$, which represent the new evidence, as well as to the standard constraints of probability theory.

New evidence reduces uncertainty. Hence, uncertainty expressed by p is, in general, smaller than uncertainty expressed by q. The principle of minimum cross-entropy helps us to determine how much it should be smaller. It allows us to reduce the uncertainty of q in the smallest amount necessary to satisfy the new evidence. That is, the posterior probability distribution function p estimated by the principle is the largest among all other distribution functions that conform to the evidence.

The principle of maximum entropy has been justified by at least three distinct arguments:

1. The maximum entropy probability distribution is the only *unbiased distribution*, that is, the distribution that takes into account all available information but no additional (unsupported) information (bias). This follows directly from the facts that

    (a) all available information (but nothing else) is required to form the constraints of the optimization problem, and

    (b) the chosen probability distribution is required to be the one that represents the maximum uncertainty (entropy) within the constrained set of probability distributions. Indeed, any reduction of uncertainty is an equal gain of information. Hence, a reduction of uncertainty from its maximum value, which would occur when any distribution other than the one with maximum entropy were chosen would mean that some information from outside the available evidence was implicitly added.

    This argument of justifying the maximum entropy principle is covered in the literature quite extensively. Perhaps its best and most thorough presentation is given in a paper by Jaynes [1979], which also contains an excellent historical survey of related developments in probability theory, and in a book by Christensen [1980-81, vol. 1].

2. It was shown by Jaynes [1968], strictly on combinatorial grounds, that the maximum probability distribution is the *most likely distribution* in any given situation.

3. It was demonstrated by Shore and Johnson [1980] that the principle of maximum entropy can be deductively derived from the following *consistency axioms* for inductive (or ampliative) reasoning:

    (ME1) Uniqueness - The result should be unique.

    (ME2) Invariance - The choice of coordinate system (permutation of variables) should not matter.

    (ME3) System independence - It should not matter whether one accounts for independent systems separately in terms of marginal probabilities or together in terms of joint probabilities.

    (ME4) Subset independence - It should not matter whether one treats an independent subset of system states in terms of separate conditional probabilities or together in terms of joint probabilities.

The rationale for choosing these axioms is expressed by Shore and Johnson as follows: any acceptable method of inference must be such that different ways of using it to take the same information into account lead to consistent results. Using the axioms, they derive the following proposition: given some information in terms of constraints regarding the probabilities to be estimated, there is only one probability distribution satisfying the constraints that can be chosen by a method that satisfies the consistency axioms; this unique distribution can be attained by maximizing the Shannon entropy (or any other function that has exactly the same maxima as the entropy) subject to the given constraints. Alternative derivations of the principle of maximum entropy were demonstrated by Smith [1974], Avgers [1983] and Paris and Vencovska [1990].

The principle of minimum cross-entropy can be justified by similar arguments. In fact, Shore and Johnson [1981] derive both principles and show that the principle of maximum entropy is a special case of the principle of minimum cross-entropy. The latter principle is further examined by Williams [1980], who shows that it generalizes the well-known Bayesian rule of conditionalization.

In general, the principles of maximum entropy and minimum cross-entropy as well as the various principles of uncertainty that extend beyond probability theory are tools for dealing with a broad class of problems referred to as *inverse problems* or *ill–posed problems* [McLaughlin, 1984; Tarantola, 1987]. A common characteristic of these problems is that they are *underdetermined* and, consequently, do not have unique solutions. The various maximum uncertainty principles allow us to obtain unique solutions to underdetermined problems by injecting uncertainty of some type to each solution to reflect the lack of information in the formulation of the problem.

As argued by Bordley [1983], the assumption that "the behavior of nature can be described in such a way as to be consistent" (so-called *consistency premise*) is essential for science to be possible. This assumption, when formulated in a particular context in terms of appropriate consistency axioms, leads to an optimization principle. According to Bordley, the various optimization principles, which are exemplified by the principles of maximum entropy and minimum cross-entropy, are central to science.

Optimization problems that emerge from the maximum uncertainty principle outside classical information theory have yet to be properly investigated and tested in praxis. When several types of uncertainty are applicable, we must choose one from several possible optimization problems. In evidence theory, for example, four distinct principles of maximum uncertainty are possible; they are distinguished from one another by the objective function involved in the associated optimization problem: nonspecificity (Eq. (3.51)), strife (Eq. (3.206)), aggregate uncertainty (Eq. (3.217)), or both nonspecificity and strife viewed as two distinct objective functions.

As a simple example to illustrate the principle of maximum nonspecificity in evidence theory, let us consider a finite universal set X, three non-empty subsets of which are of our interest: A, B, and $A \cap B$. Assume that the only evidence on hand is expressed in terms of two numbers, a and b, that represent the total beliefs focusing on A and B, respectively ($a, b \in [0, 1]$). Our aim is to estimate the degree of support for $A \cap B$ based on this evidence.

As a possible interpretation of this problem, let X be a set of diseases considered in an expert system designed for medical diagnosis in a special area of medicine, and let A and B be sets of diseases that are supported for a particular patient by some diagnostic tests to degrees a and b, respectively. Using this evidence, it is reasonable to estimate the degree of support for diseases in $A \cap B$ by using the principle of maximum nonspecificity. This principle is a safeguard that does not allow us to produce an answer (diagnosis) that is more specific than warranted by the evidence.

The use of the principle of maximum nonspecificity leads in our example to the following optimization problem:

Determine values $m(X)$, $m(A)$, $m(B)$, and $m(A \cap B)$ for which the function

$$N(m) = m(X) \log_2 |X| + m(A) \log_2 |A| + m(B) \log_2 |B| \quad (4.12)$$
$$+ m(A \cap B) \log_2 |A \cap B|$$

reaches its maximum subject to the constraints

$$m(A) + m(A \cap B) = a \qquad (4.13a)$$
$$m(B) + m(A \cap B) = b \qquad (4.13b)$$
$$m(X) + m(A) + m(B) + m(A \cap B) = 1 \qquad (4.13c)$$
$$m(X), m(A), m(B), m(A \cap B) \geq 0 \qquad (4.13d)$$

where $a, b \in [0, 1]$ are given numbers.

The constraints are represented in this case by three linear algebraic equations of four unknowns and, in addition, by the requirement that the unknowns be nonnegative real numbers. The first two equations represent our evidence; the third equation and the inequalities represent general constraints of evidence theory. The equations are consistent and independent. Hence, they involve one degree of freedom. Selecting, for example, $m(A \cap B)$ as the free variable, we readily obtain

$$m(A) = a - m(A \cap B) \qquad (4.14)$$
$$m(B) = b - m(A \cap B)$$
$$m(X) = 1 - a - b + m(A \cap B).$$

Since all the unknowns must be nonnegative, the first two equations set the upper bound of $m(A \cap B)$, whereas the third equation specifies its lower bound; the bounds are

$$\max(0, a + b - 1) \le m(A \cap B) \le \min(a, b). \tag{4.15}$$

Using Eqs. (4.14), the objective function can now be expressed solely in terms of the free variable $m(A \cap B)$. After a simple rearrangement of terms, we obtain

$$
\begin{aligned}
N(m) \;=\; & m(A \cap B) \left[\log_2 |X| - \log_2 |A| - \log_2 |B| + \log_2 |A \cap B| \right] \\
& + (1 - a - b) \log_2 |X| + a \log_2 |A| + b \log_2 |B|,
\end{aligned}
\tag{4.16}
$$

Clearly, only the first term in this expression can influence its value, so that we may rewrite the expression as

$$N(m) = m(A \cap B) \log_2 K_1 + K_2 \tag{4.17}$$

where

$$K_1 = \frac{|X| \cdot |A \cap B|}{|A| \cdot |B|} \tag{4.18}$$

and

$$K_2 = (1 - a - b) \log_2 |X| + a \log_2 |A| + b \log_2 |B| \tag{4.19}$$

are constant coefficients. The solution to the optimization problem depends only on the value of K_1. Since A, B, and $A \cap B$ are assumed to be non-empty subsets of X, $K_1 > 0$. If $K_1 < 1$, then $log_2 K_1 < 0$ and we must minimize $m(A \cap B)$ to obtain the maximum of (4.17); hence, $m(A \cap B) = \max(0, a + b - 1)$ due to (4.15). If $K_1 > 1$, then $log_2 K_1 > 0$, and we must maximize $m(A \cap B)$; hence $m(A \cap B) = \min(a, b)$ as given by (4.15). When $K = 1$, $log_2 K_1 = 0$, and $N(m)$ is independent of $m(A \cap B)$; this implies that the solution is not unique or, more precisely, that any value of $m(A \cap B)$ in the range (4.15) is a solution to the optimization problem. The complete solution can thus be expressed by the following equations:

$$
m(A \cap B) = \begin{cases}
\max(0, a + b - 1) & \text{when } K_1 < 1 \\
[\max(0, a + b - 1), \min(a, b)] & \text{when } K_1 = 1 \\
\min(a, b) & \text{when } K_1 > 1
\end{cases}
\tag{4.20}
$$

The three types of solutions are exemplified in Table 4.1.

Observe that, due to the linearity of the measure of nonspecificity, the use of the principle of maximum nonspecificity leads to linear programming problems. This is a great advantage when compared with the maximum entropy principle. The use of the latter leads to nonlinear optimization problems, which are considerably more difficult computationally.

The principle of maximum nonspecificity was first conceived by Dubois and Prade [1987c], who also demonstrated its significance in approximate reasoning based on possibility theory [Dubois and Prade, 1991]. Wierman

TABLE 4.1 Examples of the three types of solutions obtained by the principle of maximum nonspecificity. In (a) the sizes of the sets are presented. In (b) the values of a and b determine the membership grades.

(a)	$|X|$	$|A|$	$|B|$	$|A \cap B|$
$K_1 < 1$	10	5	5	2
$K_1 = 1$	10	5	4	2
$K_1 > 1$	20	10	12	7

(b)	a	b	$m(X)$	$m(A)$	$m(B)$	$m(A \cap B)$
$K_1 < 1$	0.7	0.5	0.0	0.5	0.3	0.2
$K_1 = 1$	0.8	0.6	[0,0.2]	[0.2,0.4]	[0,0.2]	[0.4,0.6]
$K_1 > 1$	0.4	0.5	0.5	0.0	0.1	0.4

[1993] has applied this principal to image processing; Padet [1996] has applied it to fuzzy control.

The use of the maximum nonspecificity principle does not always result in a unique solution, as demonstrated by the case of $K_1 = 1$ in Eq. (4.20). If the solution is not unique, then this is a good reason to utilize the second type of uncertainty – strife. We may either add the measure of strife, given by Eq. (3.206), as a second objective function in the optimization problem, or to use the measure of aggregate uncertainty, such as the sum of the non-specificity, N, and strife or discord, ST or D, as a more refined objective function. The mathematically best justified function for the maximum uncertainty principle is, of course, the aggregate uncertainty AU defined by (3.217). These variations of the maximum uncertainty principle have yet to be fully developed.

4.3   Principle of Uncertainty Invariance

Our repertory of mathematical theories by which we can characterize and deal with situations under uncertainty is already quite respectable, and it is likely that additional theories will be added to it in the future. The theories differ from one another in their meaningful interpretations, generality, computational complexity, robustness, and other aspects. Furthermore, different theories may be appropriate at different stages of a problem solving process or for different purposes at the same stage. As argued by Shackle [1961, 1979], for example, human reasoning about future events can be formalized more adequately by possibility theory rather then probability theory. Similarly, as argued by Cohen [1970], possibility theory is more suitable for formalizing inductive reasoning. Possibility theory has also some attractive properties that are very useful in reconstructability analysis of systems Klir *et al.* [1988]. On the other hand, it is well known that probability theory is a natural tool for formalizing uncertainty in situations where evidence is based on outcomes of a series of independent random experiments.

The moral is that none of the theories is superior in all respects and for all purposes. This conclusion was also reached by several studies [Henkind and Harrison, 1988; Horvitz *et al.*, 1966; Lee *et al.*, 1987; Stephanou and Sage, 1987]. The different theories of uncertainty should thus be viewed as complementary. Which of them to use in each problem situation should be decided by appropriate metarules on the basis of given problem requirements, general application context, estimated computational complexity, and other criteria. To develop such metarules, it is essential to look at the concept of uncertainty from a broad perspective (a metalevel), and use this global view to investigate connections between the various theories.

One such broad perspective, enlarging the usual semantics of propositional modal logic [Hughes and Cresswell, 1968, 1984; Chellas, 1980], was suggested by Resconi et al. [1992, 1993, 1996]. It has already been established that modal logic can indeed be used as a unifying framework for different theories of uncertainty. The completeness of modal logic representation of evidence theory was proven for finite sets [Harmanec, Klir, and Resconi, 1994] as well as infinite sets [Harmanec, Klir, and Wang, 1996; Wang, Klir, and Resconi, 1995]. It was also proven for possibility theory [Klir and Harmanec, 1994]. Modal logic was also shown to facilitate a new interpretation of fuzzy sets [Klir, 1994a]. Moreover, some new uncertainty theory, previously not considered, have emerged from the modal logic interpretation [Klir, 1994a].

In order to utilize opportunistically advantages of the various theories of uncertainty, we need the capability of moving from one theory to another as appropriate. These moves, or transformations, from one theory to another should be based on some justifiable principle. Klir [1989b, 1990a] proposed a principle for this purpose, referred to as a *principle of uncertainty invariance*. This principle, which is based on measures of uncertainty in the various theories, allows us to construct meaningful transformations between uncertainty theories.

The principle of uncertainty invariance is based upon the following epistemological and methodological position: every real-world decision or problem situation involving uncertainty can be formalized in all the theories of uncertainty. Each formalization is a mathematical model of the situation. When we commit ourselves to a particular mathematical theory, our modelling becomes necessarily limited by the constraints of the theory. For example, probability theory can model decision situations only in terms of conflicting degrees of belief in mutually exclusive alternatives. These degrees are derived in some way from the evidence on hand. Possibility theory, on the other hand, can model a decision situation only in terms of degrees of belief that are allocated to consonant (nested) subsets of alternatives; these are almost conflict-free [Geer and Klir, 1991], but involve large nonspecificity.

Clearly, a more general theory (such as evidence theory) is capable of capturing uncertainties of some decision situations more faithfully than its less general competitors (such as probability theory or possibility theory). Nevertheless, we take the position that every uncertainty theory, even the least general one, is capable of characterizing (or approximating, if you like) the uncertainty of every situation. This characterization may not be, due to the constraints of the theory, as natural as its counterparts in other, more adequate theories. We claim, however that such a characterization does always exist. If the theory is not capable of capturing some type of uncertainty directly, it may capture it indirectly in some fashion, through whatever other type of uncertainty is available.

To transform the representation of a problem solving situation in one theory, T_1, into an equivalent representation in another theory, T_2, the principle of uncertainty invariance requires that:

i. the amount of uncertainty associated with the situation be preserved when we move from T_1 into T_2; and

ii. the degrees of belief in T_1 be converted to their counterparts in T_2 by an appropriate scale, at least ordinal.

Requirement (i) guarantees that no uncertainty is added or eliminated solely by changing the mathematical theory by which a particular phenomenon is formalized. If the amount of uncertainty were not preserved then either some information not supported by the evidence would unwittingly be added by the transformation (information bias) or some useful information contained in the evidence would unwittingly be eliminated (information waste). In either case, the model obtained by the transformation could hardly be viewed as equivalent to its original.

Requirement (ii) guarantees that certain properties, which are considered essential in a given context (such as ordering or proportionality of relevant values), be preserved under the transformation. Transformations under which certain properties of a numerical variable remain invariant are known in the theory of measurement as scales.

Due to the unique connection between uncertainty and information, the principle of uncertainty invariance can also be conceived as a *principle of information invariance* or *information preservation*. Indeed, each model of a decision making situation, formalized in some mathematical theory, contains information of some type and some amount. The amount is expressed by the difference between the maximum possible uncertainty associated with the set of alternatives postulated in the situation and actual uncertainty of the model. When we approximate one model with another one, formalized in terms of a different mathematical theory, this basically means that we want to replace one type of information with an equal amount of information of another type. That is, we want to convert information from

FIGURE 4.1 Probability-possibility transformations.

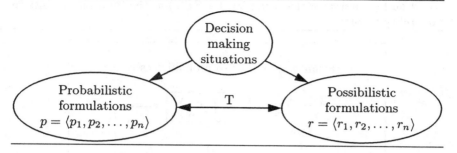

one type to another while, at the same time, preserving its amount. This expresses the spirit of the principle of information invariance or preservation: no information should be added or eliminated solely by converting one type of information to another. It seems reasonable to compare this principle, in a metaphoric way, with the principle of energy preservation in physics.

The principle of uncertainty invariance can be made operational only when the theories involved possess well-justified and unique measures of uncertainty. At this time, it is restricted to the theories whose uncertainty measures are introduced in Chapter 3. Some generic applications of the principle are discussed in the rest of this section.

Probability-Possibility Transformations

Uncertainty-invariant transformations between probabilistic and possibilistic conceptualizations of uncertainty have been investigated quite vigorously since 1989, when the principle of uncertainty invariance was first introduced. The general idea is illustrated in Fig. 4.1, where only nonzero components of the probability distribution $\mathbf{p}$ and the possibility distribution $\mathbf{r}$ are listed. It is also assumed that the corresponding components of the distributions are ordered in the same way: $p_i \geq p_{i+1}$ and $r_i \geq r_{i+1}$ for all $i = \mathbb{N}_{n-1}$. This is equivalent to the assumption that values p_i correspond to values r_i for all $i = \mathbb{N}_n$ by some scale, which must be at least ordinal.

TRANSFORMATIONS BASED ON THE SUM $U + ST$

Using the sum of the U–uncertainty (Eq. (3.38)), and strife (Eq. (3.211)) or discord (Eq. (3.210)) as total uncertainty in possibility theory, the following results regarding uncertainty-invariant transformations $\mathbf{p} \leftrightarrow \mathbf{r}$ under different scales have been obtained:

       i. transformations based on ratio and difference scales do not have enough flexibility to preserve uncertainty and, consequently, are not applicable [Klir, 1990a];

FIGURE 4.2 Uncertainty-invariant probability-possibility transformations based on log-interval scales using $U(r) + ST(r)$ as the total uncertainty in possibility theory.

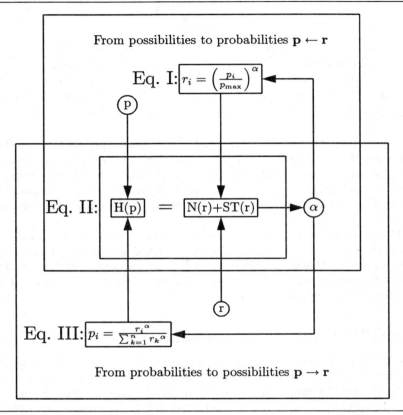

ii. for interval scales, uncertainty invariant transformations $\mathbf{p} \rightarrow \mathbf{r}$ exist and are unique for all probability distributions, while the inverse transformations $\mathbf{p} \leftarrow \mathbf{r}$ that preserve uncertainty exist (and are unique) only for some possibility distributions [Geer and Klir, 1992];

iii. for log-interval scales uncertainty-invariant transformations exist and are unique in both directions [Geer and Klir, 1992];

iv. ordinal-scale transformations that preserve uncertainty always exist in both directions, but, in general, are not unique [Klir, 1990a].

The same results were also derived, in a different way, by Jumarie [1994, 1995]. Moreover, a generalization to the continuous case in n dimensions was developed by Wonneberger [1994].

The log-interval scale is thus the strongest scale under which the uncertainty invariance transformations $\mathbf{p} \leftrightarrow \mathbf{r}$ always exist and are unique. A scheme of these transformations is shown in Fig. 4.2. First, a transformation coefficient α is determined by Eq. (II), which expresses the required equality of the two amounts of total uncertainty; then, the obtained value of α is substituted to the transformation formulas (Eq. (I) for $\mathbf{p} \rightarrow \mathbf{r}$ and Eq. (III) for $\mathbf{p} \leftarrow \mathbf{r}$. It is known that $0 < \alpha < 1$, which implies that the possibility-probability consistency condition ($r_i \geq p_i$ for all $i = 1, 2, ..., n$) is always satisfied by these transformations [Geer and Klir, 1992]. When the transformations are simplified by excluding $ST(\mathbf{r})$ in Eq. (II), which for large n is negligible [Geer and Klir, 1991], their basic properties (existence, uniqueness, consistency) remain intact.

For ordinal scales, uncertainty-invariant transformations $\mathbf{p} \leftrightarrow \mathbf{r}$ are not unique. They result, in general, in closed convex sets of probability or possibility distributions, which are obtained by solving appropriate linear inequalities constrained by the requirements of normalization and uncertainty invariance. From one point of view, the lack of uniqueness is a disadvantage of ordinal-scale transformations. From another point of view, it is an advantage since it allows us to impose additional requirements on the transformations. These additional requirements may be expressed, for example, in terms of second order properties, such as projections, noninteraction, or conditioning. Although uncertainty-invariant transformations based on ordinal scales were explored [Klir, 1990a], the utilization of their great flexibility have not been fully investigated as yet.

Examples of applications of uncertainty preserving conversions from probabilities to possibilities and vice versa can be found in the literature [Klir, 1990a, 1991c; Geer and Klir, 1992; Yoo, 1994].

The motivation to study probability–possibility transformations has arisen not only from our desire to comprehend the relationship between the two theories of uncertainty [Dubois and Prade, 1982, 1988], but also from some practical problems. Examples of these problems are: constructing a membership grade function of a fuzzy set from statistical data [Bharathi-Devi and Sarma, 1985; Civanlar and Trussell, 1986; Dubois and Prade, 1985c, 1986], combining probabilistic and possibilistic information in expert systems [Geer and Klir, 1992], constructing a probability measure from a possibility measure in the context of decision making or systems modelling [Leung, 1982; Moral, 1986; Smets, 1990a], or converting probabilities to possibilities to reduce computational complexity. To deal with these problems, various transformations have been suggested in the literature. Except for the normalization requirements, they differ from one another substantially, ranging from simple ratio scaling to more sophisticated transformations based upon various principles, such as maximum entropy [Leung, 1982; Moral, 1986], insufficient reason [Dubois and Prade, 1982, 1988; Dubois, *et al.*, 1991], maximum specificity [Dubois *et al.*, 1991], and different forms of the principle of possibility-probability consistency [Civanlar and Trussell,

1986; Delgado and Moral, 1987; Leung, 1982; Moral, 1986]. None of these transformations preserves uncertainty, but they can be made uncertainty-preserving by appropriate auxiliary scaling transformation [Klir, 1990a]. When, for example, the given transformation converts $\mathbf{p}$ to $\mathbf{r}$, the auxiliary transformation converts $\mathbf{r}$ to $\mathbf{r}'$ by the scheme in Fig. 4.2 (when we choose log-interval scaling), where p_i and p_1 in Eq. (I) are replaced with r_i and r_1, respectively, and $\mathbf{r}$ in Eq. (II) is replaced with $\mathbf{r}'$; the inverse auxiliary transformation is based on similar replacements. Clearly, the resulting $\mathbf{r}'$ preserves the uncertainty of $\mathbf{p}$ and, at the same time, some features of $\mathbf{r}$ (determined by the scale used), but it is connected to $\mathbf{p}$ only by an ordinal scale.

The various transformations that do not preserve uncertainty seem to lack a firm theoretical underpinning. The ratio-scale transformation, for example, is unnecessarily strong in preserving local properties while fully ignoring global properties such as the amount of uncertainty. Transformations based upon the principle of possibility-probability consistency seems also somewhat wanting. Although a general form of this consistency (possibilities cannot be smaller than corresponding probabilities) must be satisfied by every transformation, it is hard to see why an imposed higher level of consistency should determine the transformation. An asymmetric transformation proposed by Dubois, Prade and Sandri [1991] is also questionable. The authors argue that the possibility representation is weaker and, consequently, "turning probability into possibility measure comes down to give up part of the initial information," while "turning a possibility measure into a probability measure is always partially arbitrary since the conversion procedure always adds some information." As a result of these arguments, $\mathbf{p} \rightarrow \mathbf{r}$ is guided by the principle of maximum specificity, while $\mathbf{p} \leftarrow \mathbf{r}$ is guided by the principle of insufficient reason. These arguments may be countered as follows: Although possibility theory employs weaker rules than probability theory in manipulating uncertainty (projecting, combining, conditioning), basic structures of the two theories are not comparable. Hence, even though manipulating uncertainty within possibility theory results in a greater loss of information than corresponding manipulation of equivalent uncertainty within probability theory, it is neither necessary nor desirable to loose or gain information solely by transforming uncertainty from one representation to the other.

Consider the following experiment: Given marginal probability distributions of two interactive stochastic variables, we calculate the joint distribution $\mathbf{p}$. Transforming now the marginal probabilities into corresponding marginal possibilities, we calculate the joint possibility distribution and transform it to probability distribution $\mathbf{p}'$. Clearly, $\mathbf{p} \neq \mathbf{p}'$ due to the different ways of combining the marginal distributions in the two theories. Nevertheless, the distance between $\mathbf{p}$ and $\mathbf{p}'$, which is affected by the transformation employed, should be as small as possible. This is achieved, as demonstrated by convincing experimental results [Klir and Parviz, 1992b], by the uncertainty-preserving transformation defined in Fig. 4.2, while the mentioned asymmetric transformation [Dubois et al., 1991] for example, performs extremely poorly in this respect.

Approximations of Fuzzy Sets

A simple application of the principle of uncertainty invariance is to approximate a given fuzzy set by a crisp set that has the same nonspecificity. The resulting crisp set can then be used for defuzzification or other purposes.

Let $X = \{x_1, x_2, ..., x_n\}$ denote the universal set on which a given fuzzy set A is defined by its membership grade function $A(x_i) = a_i \in [0, 1]$. Let C denote a crisp set defined on X by its characteristic function $C(x_i) = c_i \in \{0, 1\}$. Assume that elements of X are ordered so that $a_i \geq a_{i+1}$ and $c_i \geq c_{i+1}$ for all $i = 1, 2, ..., n - 1$. Furthermore, let $\#a_i$ denote the number of repetitions of value a_i in the sequence $a_1, a_2, ..., a_n$. That is, $\#a_i = \#a_j$ when $a_i = a_j$.

We say that C is a crisp approximation of A when C is as close as possible to A and, at the same time, the nonspecificity of C is as close as possible to the nonspecificity of A.

To determine a crisp set C that approximates a given fuzzy set A is rather simple. First, we calculate the nonspecificity $U(A)$ of A by Eq. (3.31). Then, we find an integer k for which the difference $|U(A) - log_2 k|$ reaches its minimum (a trivial problem), and define C in terms of this integer:

$$C(x_i) = \begin{cases} 1 & \text{when } i \leq k \\ 0 & \text{when } i > k \end{cases} \tag{4.21}$$

When $\#a_k > 1$ and $a_{k+j} = a_k$ for some positive j, the solution is not unique. Different solutions are obtained by permuting those elements of X for which $A(x_i) = a_k$. Additional criteria (symmetry, convexity, etc.) can then be applied to choose one of the solutions.

Observe that the crisp set $C = \{6, 7, ..., 13\}$ whose characteristic function is illustrated in Fig. 3.4 by the shaded area approximates (in our sense) the fuzzy set A whose membership grade function is defined in Fig. 3.4 by the dots. Here, $U(A) = 3.05$ and, hence, $k = 8$ and $a_k = 0.6$. Since $\#a_k = 2$, there is another crisp set, $C' = \{5, 6, ..., 12\}$, which approximates A equally well. For defuzzification, it is reasonable to take, for example, the middle value of $C \cup C'$, which is 9.

Approximations in Evidence Theory

The approximation of a representation of uncertainty in evidence theory by either a probabilistic representation or a possibilistic representation is another problem area in which the principle of uncertainty invariance is directly applicable. The reason for pursuing these approximations is to reduce computational complexity.

One way of approximating an arbitrary body of evidence, $\langle \mathcal{F}, m \rangle$, by a probability measure is to distribute each value $m(A)$ uniformly over the elements of A, in the spirit of the principle of insufficient reason [Dubois and Prade, 1982, 1988; Smets, 1990a]. Another approach to probabilistic approximation was proposed by Voorbraak [1989]. Approximations of $\langle \mathcal{F}, m \rangle$ by possibility measures were recently investigated by Dubois and Prade [1990]. None of these approximations preserve uncertainty.

Uncertainty–invariant transformations from belief measures to probability measures are directly solved by calculating the value $AU(\mathrm{Bel})$ for the given belief measure Bel by Algorithm 3.1. The result of the algorithm is not only the value $AU(\mathrm{Bel})$ but also the unique probability distribution function p by which the value is calculated. This probability distribution function has the same uncertainty as the given belief measure Bel; moreover, p is consistent with Bel in the sense that

$$\sum_{x \in A} p(x) \geq \mathrm{Bel}(A) \tag{4.22}$$

for all non-empty subsets of X.

Uncertainty–invariant transformations from belief measures to necessity measures are considerably *more* complicated. To characterize these transformations, we begin with a definition of consistency of belief and necessity measures.

We say that belief measure Bel and a necessity measure Nec are consistent if they are defined on the same universal set X and

$$\mathrm{Bel}(A) \geq \mathrm{Nec}(A) \tag{4.23}$$

for all $A \in \mathcal{P}(X)$.

The motivation for this definition stems from the fact that belief functions are more general than necessity measures and therefore are, in general, more capable of expressing the evidence in hand more precisely, i.e., the interval $[\mathrm{Nec}(A), \mathrm{Pos}(A)]$ should contain the interval $[\mathrm{Bel}(A), \mathrm{Pl}(A)]$ for each $A \subseteq X$. Note that our definition of consistency is equivalent to the definition of weak inclusion of necessity measures in belief measures in the sense of Dubois and Prade [1990].

Let us assume that we are given a belief function Bel and we want to find a necessity measure Nec such that Bel and Nec are consistent and such that $AU(\mathrm{Bel}) = AU(\mathrm{Nec})$. The class of necessity measures satisfying this condition is characterized by the following theorem. A proof for this theorem as well as the proofs of the other theorems of this section are contained in a paper by Harmanec and Klir [1997].

Theorem 4.1 *Let* Bel *denote a given belief measure on* X. *Let* $\{A_i\}_{i=1}^s$, *with* $s \in \mathbb{N}_n$, *denote the partition of* X *consisting of the sets* A_i *chosen by Step 1 of Algorithm 3.1 in the ith pass through the loop of Steps 1 to*

*5 during the computation of AU(Bel). A necessity measure, Nec, that is
consistent with* Bel *satisfies the condition*

$$AU(\text{Bel}) = AU(\text{Nec}) \tag{4.24}$$

if and only if

$$\text{Nec}\left(\bigcup_{j=1}^{i} A_j\right) = \text{Bel}\left(\bigcup_{j=1}^{i} A_j\right) \tag{4.25}$$

for each $i \in \mathbb{N}_s$.

We see from Theorem 4.1 that, unless $s = n$ and $|A_i| = 1$ for all $i \in \mathbb{N}_n$, the solution to our problem is, again, not unique. The question is what criterion should be used to choose a particular solution. The answer depends on the application or problem one has at hand. Two criteria seem to be most justifiable. One criterion is to choose the necessity measure that maximizes the nonspecificity among all the necessity measures that satisfy the conditions of Theorem 4.1. The second criterion is to pick out a necessity measure that is in some sense the closest to the belief measure from all acceptable necessity measures. The following theorem deals with the first criterion.

Theorem 4.2 *Under the notation of Theorem 4.1, the necessity measure, Nec, that maximizes the nonspecificity among the necessity measures consistent with* Bel *and containing the same amount of uncertainty is given by*

$$\text{Nec}(B) = \text{Bel}\left(\bigcup_{j=1}^{i} A_j\right) \tag{4.26}$$

for all $B \subseteq X$, *where* i *is the largest integer such that* $\bigcup_{j=1}^{i} A_j \subseteq B$. *(We assume that* $\bigcup_{j=1}^{0} A_j = \emptyset$ *by convention.)*

In order to pursue the second criterion, we need a suitable way of measuring the closeness of a necessity measure to a given belief function. For this purpose, we introduce a function DF_{Bel} defined by

$$DF_{\text{Bel}}(\text{Nec}) = \sum_{A \in \mathcal{P}(X)} (\text{Bel}(A) - \text{Nec}(A)) \tag{4.27}$$

for any necessity measure Nec defined on X and consistent with Bel. We minimize this function over all necessity measures consistent with Bel and containing the same amount of uncertainty. That is, we try to minimize the sum of differences of a given belief function and the "approximating" necessity measure. The following theorem reduces the class of necessity measures we need to examine.

Theorem 4.3 *Let* Bel *be a given belief function on* X. *Under the notation employed in Theorem 4.1, the necessity measure,* Nec, *that minimizes the function* DF_{Bel}, *among all necessity measures that are consistent with* Bel *and contain the same amount of uncertainty as* Bel, *satisfies the equation*

$$\text{Nec}\left(\{x_1, x_2, ..., x_l\}\right) = \text{Bel}\left(\{x_1, x_2, ..., x_l\}\right) \tag{4.28}$$

for each $l \in \mathbb{N}_n$ *and some ordering* $x_1, x_2, ..., x_n$ *of* X, *such that each focal element of* Nec *has the form* $\{x_1, x_2, ..., x_l\}$ *for some* $l \in \mathbb{N}_n$ *and for each* $i \in \mathbb{N}_s$ *there is some* l_i *satisfying* $\{x_1, x_2, ..., x_{l_i}\} = \bigcup_{j=1}^{i} A_j$ *(i.e. the sets* $\bigcup_{j=1}^{i} A_j$ *remain fixed).*

We can express DF_{Bel} as follows

$$DF_{\text{Bel}}(\text{Nec}) = \sum_{A \in \mathcal{P}(X)} (\text{Bel}(A) - \text{Nec}(A)) \tag{4.29}$$

$$= \sum_{A \in \mathcal{P}(X)} \text{Bel}(A) - \sum_{A \in \mathcal{P}(X)} \text{Nec}(A).$$

Since $\sum_{A \in \mathcal{P}(X)} \text{Bel}(A)$, $\text{Nec}(\emptyset)$, and $\text{Nec}(X)$ are constants, minimizing DF_{Bel} is equivalent to maximizing the expression

$$\sum_{A \in \mathcal{P}(X) - \{\emptyset, X\}} \text{Nec}(A) \tag{4.30}$$

over all necessity measures Nec that are consistent with Bel and satisfy $AU(\text{Bel}) = AU(\text{Nec})$. It follows from Theorem 4.3 that maximizing the expression (4.30) is in turn equivalent to maximizing the expression

$$\sum_{k=1}^{n-1} \sum_{A \in \Pi_k} \text{Bel}\left(\{x_{\pi(1)}, x_{\pi(2)}, ..., x_{\pi(k)}\}\right) \tag{4.31}$$

$$= \sum_{k=1}^{n-1} 2^{n-1-k} \text{Bel}\left(\{x_{\pi(1)}, x_{\pi(2)}, ..., x_{\pi(k)}\}\right)$$

over all permutations π of $\mathbb{N}_n$ such that for each $i \in \mathbb{N}_s$ there is an l_i satisfying $\{x_{\pi(1)}, x_{\pi(2)}, ..., x_{\pi(l_i)}\} = \bigcup_{j=1}^{i} A_j$, and where

$$\Pi_k = \left\{ A \in \mathcal{P}(X) - \{\emptyset, X\} \mid \begin{array}{l} \{x_{\pi(1)}, x_{\pi(2)}, ..., x_{\pi(k)}\} \subseteq A \\ \{x_{\pi(1)}, x_{\pi(2)}, ..., x_{\pi(k+1)}\} \not\subseteq A \end{array} \right\}, \tag{4.32}$$

and s and $\{A_j\}_{j=1}^{s}$ have the same meaning as in Theorem 4.1. This suggests one possible way to solve our problem: Compute the value (4.31) for each permutation π satisfying the above requirements and take any one with the maximal value of (4.31); a sought necessity measure is uniquely determined by the values $\{x_{\pi(1)}, x_{\pi(2)}, ..., x_{\pi(k)}\}$, $k \in \mathbb{N}_{n-1}$.

An efficient algorithm for this computation is presented and justified in [Harmanec and Klir, 1997].

Revised Probability-Possibility Transformations

The results regarding uncertainty invariant transformations between probabilities and possibilities as described previously are somewhat compromised. This is caused by the fact that possibilistic measure of uncertainty upon which they are based is not subadditive. It was thus considered important to reinvestigate the transformations in terms of the well–justified aggregate uncertainty measure AU [Harmanec and Klir, 1997]. A summary of results obtained by this reinvestigation are presented in this section.

Let us now assume that we are given a possibility distribution $\mathbf{r} = \langle r_1, r_2, ..., r_n \rangle$. We want to find a probability distribution $\mathbf{p} = \langle p_1, p_2, ..., p_n \rangle$ such that the amounts of uncertainty contained in $\mathbf{p}$ and $\mathbf{r}$ are the same, and such that $\mathbf{p}$ is consistent with $\mathbf{r}$. Possibility distribution $\mathbf{r}$ and probability distribution $\mathbf{p}$ are said to be *consistent* if it holds that for any $A \in X$

$$Pro(A) \leq Pos(A) \tag{4.33}$$

where Pro denotes the probability measure corresponding to $\mathbf{p}$ and Pos denotes the possibility measure corresponding to $\mathbf{r}$. The justification for this requirement can be summarized by saying that what is probable is also possible. It is easy to see that (4.33) holds iff

$$\sum_{i=k}^{n} p_i \leq r_k \tag{4.34}$$

for all $k \in \mathbb{N}_n$

Examining Algorithm 3.2 we realize that it gives not only the value of $AU(Pos)$, but also the maximizing probability distribution $\mathbf{p}$ from among all probability distribution consistent with $\mathbf{r}$. It follows from the concavity of the Shannon entropy that this probability distribution is unique. The following theorem guarantees that the probability distribution $\mathbf{p}$ obtained by Algorithm 3.2 is ordered as required.

Theorem 4.4 *Let* $\mathbf{r} = \langle r_1, r_2, ..., r_n \rangle$ *denote a given ordered possibility distribution and let* $\mathbf{p} = \langle p_1, p_2, ..., p_n \rangle$ *denote the probability distribution obtained by Algorithm 3.2. Then* $p_k \geq p_{k+1}$ *for all* $k \in \mathbb{N}_{n-1}$.

Proof. Fix arbitrary $k \in \mathbb{N}_{n-1}$. We know from Algorithm 3.2 that there are $j, i \in \mathbb{N}_n$ such that $j \leq k \leq i$ and $\frac{r_i - r_{i+1}}{i+1-j} = p_k$. If $i \geq k + 1$ then $p_k = p_{k+1}$ and the theorem holds. Assume now that $i = k$. Then there exists i' such that $k + 1 \leq i'$ and $\frac{r_{k+1} - r_{i'+1}}{(i'+1)-(k+1)} = p_{k+1}$. We know by the assumption that $i = k$ in Algorithm 3.2 that

$$\frac{r_j - r_{k+1}}{(k+1) - j} > \frac{r_j - r_{i'+1}}{(i'+1) - j}. \tag{4.35}$$

It follows from (4.35) that

$$\frac{(i'+1) - j}{(k+1) - j} \left(r_j - r_{k+1} \right) > r_j - r_{i'+1}, \tag{4.36a}$$

$$\frac{[(i'+1)-(k+1)]+[(k+1)-j]}{(k+1)-j}(r_j-r_{k+1})>r_j-r_{i'+1}, \quad (4.36b)$$

$$\frac{[(i'+1)-(k+1)]}{(k+1)-j}(r_j-r_{k+1})>(r_j-r_{i'+1})-(r_j-r_{k+1}), (4.36c)$$

$$\frac{[(i'+1)-(k+1)]}{(k+1)-j}(r_j-r_{k+1})>(r_{k+1}-r_{i'+1}), \quad\quad (4.36d)$$

and

$$\frac{r_j-r_{k+1}}{(k+1)-j}>\frac{r_{k+1}-r_{i'+1}}{(i'+1)-(k+1)}. \quad\quad (4.37)$$

But the last inequality means that $p_k > p_{k+1}$, which concludes the proof of the theorem.   ∎

Example 4.2 *Let $X = \{1,2,3\}$ and $\mathbf{r} = \langle 1.0, 0.7, 0.3 \rangle$. In the first loop of Algorithm 3.2 we get $i = 2$ and $p_1 = p_2 = 0.35$. The result of the second loop is $p_3 = 0.3$ (and $AU(Pos) = -0.7 \log_2 0.35 - 0.3 \log_2 0.3 \approx 1.581$). Therefore the uncertainty preserving transformation of $\mathbf{r}$ is $\mathbf{p} = \langle 0.35, 0.35, 0.3 \rangle$.*

Example 4.2 shows that the uncertainty–preserving transformation from possibility distributions to probability distribution preserves ordering, but it does not preserve strict ordering ($r_1 > r_2$ but $p_1 = p_2$).

Let us now investigate the inverse problem. That is, we assume that a probability distribution $\mathbf{p}$ is given, and we want to find a possibility distribution $\mathbf{r}$, such that $\mathbf{r}$ is consistent with $\mathbf{p}$, and such that the amounts of uncertainty captured by $\mathbf{p}$ and $\mathbf{r}$ is equal. Let us start with an example.

Example 4.3 *Let $X = \{1,2,3\}$ and $\mathbf{p} = \langle 0.35, 0.35, 0.3 \rangle$. The result of Algorithm 3.2 is $\mathbf{r} = \langle 1.0, 1.0, 0.3 \rangle$ with $AU(Pos) \approx 1.581$.*

From Examples 4.2 and 4.3 we see that for a given probability distribution $\mathbf{p}$ there are many possibility distributions that are consistent with $\mathbf{p}$ and contain the same amount of uncertainty as $\mathbf{p}$. Therefore, the solution to the above formulated problem may not be unique. To make the solution unique we require, in addition, that $\mathbf{p}$ and $\mathbf{r}$ be equivalent in the ordinal sense, i.e., $p_i \geq p_j$ iff $r_i \geq r_j$ for all $i, j \in \mathbb{N}_n$. (Now the possibility distribution $\mathbf{r}$ from Example 4.3 becomes the unique solution for $\mathbf{p} = \langle 0.35, 0.35, 0.3 \rangle$.)

Assume that a probability distribution $\mathbf{p} = \langle p_1, p_2, ..., p_n \rangle$ is given. Let k denote the number of distinct values in the n–tuple $\langle p_1, p_2, ..., p_n \rangle$. Then there exist k integers, $s_1, s_2, ..., s_k \in \mathbb{N}_n$ such that $s_1 < s_2 < ... < s_k \leq n$ and $p_1 = p_2 = p_{s_1} > p_{s_1+1} = p_{s_1+2} = ... = p_{s_2} > p_{s_2+1} \cdots p_{s_k-1} > p_{s_{k-1}+1} = p_{s_{k-1}+2} = ... = p_{s_k}$. It follows from the requirement of ordinal equivalence between $\mathbf{p}$ and $\mathbf{r}$ that the resulting possibility distributions $\mathbf{r}$ also has k distinct values, and $r_1 = r_2 = r_{s_1} > r_{s_1+1} = r_{s_1+2} = ... = r_{s_2} > r_{s_2+1} \cdots r_{s_{k-1}} > r_{s_{k-1}+1} = r_{s_{k-1}+2} = ... = r_{s_k}$. We can now show by induction on j

that the unique solution to our problem is

$$r_{s_j} = \sum_{i=s_{j-1}+1}^{n} p_i \qquad (4.38)$$

for $j \in \mathbb{N}_n$ with $s_0 = 0$. From the normalization requirement, we have $r_{s_1} = r_1 = 1 = \sum_{i=1}^{n} p_i$. Assume $r_{s_j} = \sum_{i=s_{j-1}+1}^{n} p_i$. By examining Algorithm 3.2, we obtain

$$\frac{r_{s_{j-1}+1} - r_{s_j+1}}{(s_j + 1) - (s_{j-1} + 1)} = p_{s_j} . \qquad (4.39)$$

From (4.39), we have

$$r_{s_j} - r_{s_{j+1}} = r_{s_{j-1}+1} - r_{s_j+1} = (s_j - s_{j-1}) p_{s_j} = \sum_{i=s_{j-1}+1}^{s_j} p_i , \qquad (4.40)$$

which implies

$$r_{s_{j+1}} = r_{s_j} - \sum_{i=s_{j-1}+1}^{s_j} p_i = \sum_{i=s_{j-1}+1}^{n} p_i - \sum_{i=s_{j-1}+1}^{s_j} p_i = \sum_{i=s_j+1}^{n} p_i . \qquad (4.41)$$

Our statement then follows by induction on j.

Let us conclude this subsection with a numerical example illustrating the discussed transformation.

Example 4.4 *Let $X = \{1, 2, 3, 4, 5, 6, 7, 8\}$. Assume we are given the probability distribution $\mathbf{p} = \langle 0.3, 0.2, 0.2, 0.1, 0.08, 0.04, 0.04, 0.04 \rangle$ and we want to convert it into a possibility distribution $\mathbf{r}$ that preserves the uncertainty of $\mathbf{p}$. Using (4.38) we get*

$$\mathbf{p} = \langle 1.0, 0.7, 0.7, 0.3, 0.2, 0.12, 0.12, 0.12 \rangle . \qquad (4.42)$$

For example $r_3 = \sum_{i=2}^{8} p_i = 0.2 + 0.2 + 0.1 + 0.08 + 0.04 + 0.04 + 0.04 = 0.7$.

4.4 Summary of Uncertainty Principles

Let us summarize now, with the help of Fig. 4.3, basic ideas of the three uncertainty principles. Given an experimental frame within which we choose to operate, and some evidence regarding entities of this frame, we express information based upon the evidence in terms of some theory of uncertainty. This expression may be viewed as our *information domain* within which we can deal with a variety of associated problems. Four classes of problems are readily distinguished:

FIGURE 4.3 Overview of uncertainty principles: reduction — principle of minimum uncetainty; extension — principle of maximum uncertainty; invariant transformation — principle of uncertainty invariance.

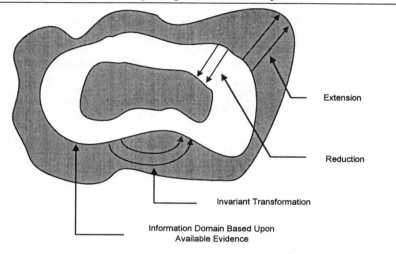

Extension

Reduction

Invariant Transformation

Information Domain Based Upon
Available Evidence

1. Problems that involve *neither a change in information domain nor a change of the uncertainty theory employed.* These are problems that can be solved by appropriate chains of logical inference within the information domain. It is required that logical inference be used correctly, but no additional principles, external to logic, are needed.

2. Problems that involve a *reduction of the information domain* and consequently, information loss. As a rule, these problems result in multiple solutions. A principle is needed by which we choose meaningful solutions from a given solution set. Since the loss of information is generally undesirable, at least on epistemological grounds, the principle of minimum uncertainty is pertinent; it ensures that we solve the problem with minimum information loss.

3. Problems that involve an *extension of the information domain.* Information contained in the information domain *is* not sufficient to determine unique solutions to these problems. To choose meaningful solutions from all solutions compatible with the information domain, we need to resort to a principle. Since any introduction of information from outside the information domain (i.e., information not supported by evidence) is unwarranted on epistemological grounds, we need to avoid it. This can be accomplished by the maximum uncertainty principle. By maximizing uncertainty subject to the constraints expressed by the given information domain, a solution is chosen (or constructed)

whose amount of information does not exceed the amount of information associated with the information domain.

4. Problems in which the *information domain expressed in one theory is transformed into a corresponding expression in another theory*. While information is represented differently in the two theories, no information should be unwittingly added or eliminated solely by changing the representation. The uncertainty invariance principle ensures that the amount of information expressed by the evidence remains constant no matter in which form it is expressed.

5
CONCLUSIONS

5.1 Appraisal of Current Results

A turning point in our understanding of the concept of uncertainty was reached when it became clear that there are several types of uncertainty. This new insight was obtained by examining uncertainty emerging from mathematical theories more general than classical set theory and probability theory.

We recognize now that the Hartley function and the Shannon entropy measure distinct types of uncertainty, nonspecificity and strife, respectively. This distinction was concealed within the confines of classical information theory, in which the Hartley function is almost routinely viewed as a special case of the Shannon entropy, emerging for uniform probability distributions. This view, which is likely a result of the fact that the value of the Shannon entropy of the uniform probability distribution on some set is equal to the value of the Hartley function for the same set, was ill-conceived. Indeed, the Hartley function is totally independent of any probabilistic assumptions, as correctly recognized by Kolmogorov [1965], who refers to the Hartley function as "the combinatorial approach to the quantitative definition of information" and makes a relevant remark: "Discussions of information theory do not usually go into this combinatorial approach at any length, but I consider it important to emphasize its logical independence of probabilistic assumptions." By and large, Kolmogorov's opinion has not influenced information theorists who continue to this day to subsume the Hartley function under the Shannon entropy or to dismiss it altogether. One rare exception was Rényi [1970], who developed an axiomatic characterization of the Hartley function and proved its uniqueness.

Since the possibility or impossibility of each relevant alternative is the only feature to which the Hartley function is sensitive, the function can be generalized only by allowing degrees of possibilities. This avenue opened with the emergence of possibility theory [Zadeh, 1978]. When possibilistic U-uncertainty was firmly established [Higashi and Klir, 1983a], the significance of the Hartley function was asserted. By recognizing that the U-uncertainty is an expected value of the Hartley function for focal elements (or α-cuts), its generalization to evidence theory became rather obvious.

Results obtained for the generalizations of the Hartley measure in possibility theory, fuzzy set theory, evidence theory, and fuzzified evidence theory, which measure nonspecificity in its various forms, are perhaps the most significant results regarding nonprobabilistic uncertainty-based information. The results, obtained primarily in the 1980s, are strong and quite conclusive.

Research regarding the generalization of the Shannon entropy in evidence theory has been more complicated. Although it was recognized already in the early 1980s [Yager, 1983] that the Shannon entropy measures average conflict among evidential claims, it was not obvious how to formulate entropy-like measure in evidence theory. Several candidates have been proposed over the years. At one time the function ST defined by Eq. (3.206) and referred to as a measure of strife appeared to be the best choice. Although this function was well justified on intuitive grounds, its lack of subadditivity was particularly troublesome. The newly proposed measure of aggregate uncertainty AU, Eq. (3.217), now appears to be the best candidate. It possesses all the desirable properties as described in Sec. 3.3. Its emergence is too recent for conclusive results, and an axiomatic characterization from appropriate axioms is needed, but its prospects are very good. It is interesting to note the connection between the definition of the measure AU and the maximum-entropy methods described in Sec. 4.2.

Literature dealing with measures of fuzziness is quite extensive. Several approaches to measuring fuzziness have been proposed and studied. The view that the degree of fuzziness of a fuzzy set is best expressed in terms of the lack of distinction between the set and its complement seems predominant. The advantage of this view is that it takes into account different definitions of fuzzy complement [Higashi and Klir, 1982] and, indirectly, different definitions of fuzzy intersection and fuzzy union [Klir and Folger, 1988]. In these general terms, the degree of fuzziness of a fuzzy set is expressed in terms of the lack of distinction between the union and intersection of the set and its complement. Using this formulation, Eq. (3.258) assumes a generalized form

$$f(A) = \sum_{x \in X} \left[1 - u\left(A(x), A^{co}(x)\right) + i\left(A(x), A^{co}(x)\right)\right] \qquad (5.1)$$

where A^{co} denotes an arbitrary fuzzy complement and u, i denote, respectively, an arbitrary fuzzy union and an arbitrary fuzzy intersection, usually t-norms and t-conorms, respectively.

The various novel principles of minimum uncertainty and maximum uncertainty have not been properly developed as yet. When developed, they will undoubtedly reach far beyond the proven utility of the rather restricted principles of minimum entropy and maximum entropy. The most promising in the immediate future is the principle of maximum nonspecificity, which has already been tested quite successfully in the area of image processing [Wierman, 1993], and the principles based on the aggregate uncertainty AU.

The principle of uncertainty invariance may be viewed as a metaprinciple. It provides us with a rational framework for constructing sensible transformations between the various uncertainty theories. As a consequence, it helps us to view these theories from a broader, unified perspective. Thus far the principle has primarily been applied to the study of transformations between probability theory, possibility theory, and evidence theory. Significant results, both theoretical [Geer and Klir, 1992; Harmanec and Klir, 1996, 1997] and experimental [Klir and Parviz, 1992b], regarding these transformations were obtained.

Probability theory has for long time been considered the only mathematical framework capable of describing uncertainty. This view is still maintained by some probabilists [Klir, 1989a, 1994b]. The growing inventory of results pertaining to uncertainty liberated from its probabilistic confines, as summarized in this book, demonstrate that this view is increasingly untenable. Although these results are quite impressive, they represent, in our opinion, only a beginning of a new, broad, and important area of research. In the rest of this chapter, we overview some open problems of immediate interest that emerge from this new area, and outline future directions of research this area is likely to undertake.

5.2   Unresolved Problems

Of all the nonclassical uncertainty measures introduced in Chapter 3, the best understood and justified are the measures of nonspecificity, functions N and U. Although these functions are known to possess surprising number of important properties [Klir and Wierman, 1987], their further investigation seems still worthwhile. As discovered by Ramer [1987], for example, function N can be interpreted as the value of the first derivative of the logarithm of the Dirichlet function at zero. Exploring this and other possible avenues is likely to further advance our understanding of this delicate function.

The relative measure of nonspecificity applicable to continuous possibility distributions, which is expressed by Eq. (3.51), still remains to be properly investigated from both theoretical and practical points of view. Its generalized counterpart in evidence theory has not even been contemplated thus far. Hence, a host of problems regarding relative measures of nonspecificity for bodies of evidence defined on infinite universal sets remain unresolved at this time.

There are even more unresolved problems associated with the measure of aggregate uncertainty AU. Although this measure emerged as the best justified candidate for an uncertainty measure in evidence theory, we have not developed its sound axiomatic characterization as yet. This is a serious theoretical weakness, which makes this measure of aggregate uncertainty less justified than the measure of nonspecificity. Moreover, we need to formulate relative measure of aggregate uncertainty for both finite and infinite sets.

A comparative study of the various measures of fuzziness based on sensible requirements is needed to make convincing choices. For example, the requirement that the degree of fuzziness be dependent on the operators of fuzzy complementation, union, and intersection employed in each particular application would favor the measure defined by Eq. (5.1) over most other measures of fuzziness. The question of uniqueness should eventually be addressed as well.

Measures of fuzziness defined for discrete membership grade functions, such as those expressed by Eq. (3.258) or Eq. (3.262), can readily be modified for continuous membership grade functions by replacing the summation in the formulas involved with integration. Although both types of measures satisfy basic requirements for measures of fuzziness, the connection between them is not obvious and should be investigated.

Each fuzzy set has some degree of nonspecificity and some degree of fuzziness. Is it reasonable to consider that each fuzzy set has also some degree of strife among its α-cuts? Formally, this consideration makes sense since the possibilistic measure of strife is readily applicable to fuzzy sets. However, it raises an interesting question: Is the strife expressing another type of uncertainty associated with fuzzy sets, or is it rather another measure of fuzziness? This question is still unresolved.

A host of unresolved problems are connected with the three principles of uncertainty overviewed in Chapter 4. One group of problems involves justification of the various alternatives of the principle of maximum uncertainty (maximum nonspecificity, maximum total uncertainty, etc.) Ideally, these principles should be derived from appropriate consistency axioms á la Shore and Johnson [1980]. Another direction of research involves the formulation and algorithmic development of the associated optimization problems for the various objective functions and representative types of constraints.

5.3   Future Directions

The various unresolved problems mentioned in the previous section are closely connected with past developments in uncertainty-based information. The purpose of this section is to look ahead to prospective future developments in this area.

Thus far, measures of uncertainty are available only in five theories that are capable of formalizing uncertainty: classical set theory, fuzzy set theory, possibility theory, probability theory, and evidence theory. There are other theories of uncertainty in which the issue of quantifying uncertainty has not been investigated as yet. Examples of these theories are: a theory based on convex sets of probability distributions [Kyburg, 1987] and other theories of imprecise probabilities [Walley and Fine, 1979; Walley, 1991], theory of rough sets [Pawlak, 1991], and theories based on some special types of fuzzy measures, such as Sugeno λ-measures or autocontinuous fuzzy measures, as well as the full fuzzy measure theory [Wang and Klir, 1992]. A search for adequate measures of uncertainty in these various theories will undoubtedly be one main direction of research in the area of uncertainty-based information.

Our broadening view of the concept of uncertainty and uncertainty-based information has opened new possibilities for studying higher aspects of information. As shown by Dretske [1981, 1983], a study of semantic aspects of information requires a well founded underlying theory of uncertainty-based information. While Dretske relies in his study only on information expressed in terms of the Shannon entropy, the broader view of uncertainty-based information allows us now to approach the same study in a more flexible and, consequently, more meaningful fashion. Studies regarding pragmatic aspects of information, such as those by Whittemore and Yovits [1973, 1974] and Yovits *et al.* [1981] should be affected likewise. Hence, the use of the various novel uncertainty theories, uncertainty measures, and uncertainty principles in the study of semantic and pragmatic aspects of information will likely be another main direction of research.

All the recent developments regarding uncertainty-based information result from the generalization of classical set theory into fuzzy set theory and from the generalization of classical measure theory into fuzzy measure theory. It is interesting to observe that generalizing mathematical theories is a current trend in mathematics. In addition to the two mentioned generalizations that are relevant to this chapter, we can easily find other examples of generalizations:

- — from quantitative theories to *qualitative theories;*

- — from functions to *relations;*

- — from graphs to *hypergraphs;*

- — from ordinary geometry (Euclidean as well as non-Euclidean) to *fractal geometry*;

- — from ordinary automata to *dynamic cellular automata*;

- — from linear theories to *nonlinear theories*;

- — from regularities to singularities (*catastrophe theory*);

- — from precise analysis to *interval analysis*;

- — from regular languages to *developmental languages*;

- — from intolerance to inconsistencies of all kinds to the *logic of inconsistency*;

- — from single objective criteria optimization to *multiple objective criteria optimization*;

Each generalization of a mathematical theory usually results in a conceptually simpler theory. This is a consequence of the fact that some properties of the former theory are not required in the latter. At the same time, the more general theory has always a greater expressive power, which, however, is achieved only at the cost of greater computational demands. This explains why these generalizations are closely connected with advances in computer technology and associated increases in computing power.

It seems appropriate to conclude this chapter by the recognition that the various generalizations in mathematics have enriched not only our insights but, together with computer technology, extended also our capabilities for modelling the intricacies of the real world.

A

UNIQUENESS OF THE U-UNCERTAINTY

$1_m \Rightarrow$ seq of 1's repeated m X
$0_m \Rightarrow$ seq of 0's ―ic

need 5 requirements (axioms)

In the following lemmas, function U is assumed to satisfy axioms (U1), (U3), (U6), and (U9) presented in Sec. 3.1 (pp. 52-52). The notation $1_m = 1,1,1,\ldots,1$ and $0_m = 0,0,0,\ldots,0$ is also carried over from the text.

shows from branch axiom can develop a more

Lemma A.1 *For all $q, k \in \mathbb{N}$, $2 \le q < k \le n$, with $\mathbf{r} = \langle r_1, r_2, \ldots, r_n \rangle$*

gen'l branching, now not just skipping 1 level but $q-1$
levels for $q \ge 2$

$$U(\mathbf{r}) = U(r_1, r_2, \ldots, r_{k-q}, \underbrace{r_k, r_k, \ldots, r_k}_{q}, r_{k+1}, \ldots, r_n) \quad (A.1)$$

$q = 2 = $ reg. branching

$$+(r_{k-q} - r_k)U(1_{k-q}, \frac{r_{k-q+1} - r_k}{r_{k-q} - r_k}, \frac{r_{k-q+2} - r_k}{r_{k-q} - r_k},$$

$$\ldots, \frac{r_{k-1} - r_k}{r_{k-q} - r_k}, 0_{n-k+1})$$

$$-(r_{k-q} - r_k)U(1_{k-q}, 0_{n-k+q})$$

Proof. By the branching axiom we have *use math. induction; holds*

for some m + holds for more

$$U(\mathbf{r}) = U(r_1, r_2, \ldots, r_{k-2}, r_k, r_k, r_{k+1}, \ldots, r_n) \quad (A.2)$$

$$+(r_{k-2} - r_k)U(1_{k-2}, \frac{r_{k-1} - r_k}{r_{k-2} - r_k}, 0_{n-k+1})$$

$$-(r_{k-2} - r_k)U(1_{k-2}, 0_{n-k+2})$$

This also shows that the lemma is true for $q = 2$. We now proceed with a proof by induction on q. Applying the induction hypothesis to the first term on the right hand of Eq. (A.2), we get

$$U(r_1, r_2, ..., r_{k-2}, r_k, r_k, r_{k+1}, ..., r_n) \tag{A.3}$$
$$= U(r_1, r_2, ..., r_{k-q}, \underbrace{r_k, r_k, ..., r_k}_{q-1}, r_k, r_{k+1}, ..., r_n)$$
$$+(r_{k-q} - r_k)U(\mathbf{1}_{k-q}, \frac{r_{k-q+1} - r_k}{r_{k-q} - r_k}, \frac{r_{k-q+2} - r_k}{r_{k-q} - r_k},$$
$$..., \frac{r_{k-2} - r_k}{r_{k-q} - r_k}, \mathbf{0}_{n-k+2})$$
$$-(r_{k-q} - r_k)U(\mathbf{1}_{k-q}, \mathbf{0}_{n-k+q})$$

If we substitute this back into Eq. (A.2), we deduce that

$$U(\mathbf{r}) = U(r_1, r_2, ..., r_{k-q}, \underbrace{r_k, r_k, ..., r_k}_{q}, r_{k+1}, ..., r_n) \tag{A.4}$$
$$+(r_{k-q} - r_k)U(\mathbf{1}_{k-q}, \frac{r_{k-q+1} - r_k}{r_{k-q} - r_k}, \frac{r_{k-q+2} - r_k}{r_{k-q} - r_k},$$
$$..., \frac{r_{k-2} - r_k}{r_{k-q} - r_k}, \mathbf{0}_{n-k+2})$$
$$-(r_{k-q} - r_k)U(\mathbf{1}_{k-q}, \mathbf{0}_{n-k+q})$$
$$+(r_{k-2} - r_k)U(\mathbf{1}_{k-2}, \frac{r_{k-1} - r_k}{r_{k-2} - r_k}, \mathbf{0}_{n-k+1})$$
$$-(r_{k-2} - r_k)U(\mathbf{1}_{k-2}, \mathbf{0}_{n-k+2})$$

Now apply the branching axiom to the quantity

$$(r_{k-q} - r_k)U(\mathbf{1}_{k-q}, \frac{r_{k-q+1} - r_k}{r_{k-q} - r_k}, \frac{r_{k-q+2} - r_k}{r_{k-q} - r_k}, \tag{A.5}$$
$$..., \frac{r_{k-1} - r_k}{r_{k-q} - r_k}, \mathbf{0}_{n-k+1})$$

to produce

$$(r_{k-q} - r_k)U(\mathbf{1}_{k-q}, \frac{r_{k-q+1} - r_k}{r_{k-q} - r_k}, \frac{r_{k-q+2} - r_k}{r_{k-q} - r_k}, \tag{A.6}$$
$$..., \frac{r_{k-1} - r_k}{r_{k-q} - r_k}, \mathbf{0}_{n-k+1})$$
$$= (r_{k-q} - r_k)U(\mathbf{1}_{k-q}, \frac{r_{k-q+1} - r_{k-1}}{r_{k-q} - r_k}, \frac{r_{k-q+1} - r_k}{r_{k-q} - r_k},$$
$$..., \frac{r_{k-2} - r_k}{r_{k-q} - r_k}, \mathbf{0}_{n-k+2})$$
$$+(r_{k-2} - r_k)U(\mathbf{1}_{k-2}, \frac{r_{k-1} - r_k}{r_{k-2} - r_k}, \mathbf{0}_{n-k+1})$$
$$-(r_{k-2} - r_k)U(\mathbf{1}_{k-2}, \mathbf{0}_{n-k+2})$$

All three terms on the right hand side of the equality in Eq. (A.6) are present in the right hand side of Eq. (A.3) so that by a simple substitution we conclude that which was to be proved:

$$U(\mathbf{r}) = U(r_1, r_2, ..., r_{k-q}, \underbrace{r_k, r_k, ..., r_k}_{q}, r_{k+1}, ..., r_n) \qquad (A.7)$$

$$+(r_{k-q} - r_k)U(\mathbf{1}_{k-q}, \frac{r_{k-q+1} - r_k}{r_{k-q} - r_k}, \frac{r_{k-q+2} - r_k}{r_{k-q} - r_k},$$

$$..., \frac{r_{k-1} - r_k}{r_{k-q} - r_k}, \mathbf{0}_{n-k+1})$$

$$-(r_{k-q} - r_k)U(\mathbf{1}_{k-q}, \mathbf{0}_{n-k+q}).$$

identical to Hartley proof, " define g(k)

Lemma A.2 *For $k \in \mathbb{N}$, $k < n$, define $g(k) = U(\mathbf{1}_k, \mathbf{0}_{n-k})$, then $g(k) = \log_2 k$.*

apply H(x) to a crispset of k elements = log k just like

Proof. By the expansibility axiom we know that $U(\mathbf{1}_k, \mathbf{0}_{n-k}) = U(\mathbf{1}_k)$. By the additivity axiom $g(k + l) = g(k) + g(l)$ so by a proof identical to that of Theorem 3.1 we can conclude that $g(k) = \log_2 k$.

proving Hartley measure

Lemma A.3 *For $k \in \mathbb{N}$, $k \geq 2$, and $\rho \in \mathbb{R}$, $0 \leq \rho \leq 1$ we have*

$$U(\mathbf{1}_k, \boldsymbol{\rho}_{n-k}) = (1 - \rho)\log_2 k + \rho \log_2 n \qquad (A.8)$$

where $\boldsymbol{\rho}_{n-k}$ represents $\underbrace{\rho, \rho, ..., \rho}_{n-k}$.

need unit, expan, add. to prepare for H. proof

Proof. Let $\mathbf{r}$ be the possibility distribution $\mathbf{1}_k$, $\boldsymbol{\rho}_{n-k}$. Form the joint possibility distribution $\mathbf{r}^2$ where both marginal distributions are equivalent to $\mathbf{r}$. It is simple to check that $\mathbf{r}^2 = \mathbf{1}_{k^2}$, $\boldsymbol{\rho}_{n^2-k^2}$ has this property. Now $U(\mathbf{r}^2) = U(\mathbf{1}_{k^2}, \boldsymbol{\rho}_{n^2-k^2})$. First we can deduce from the additivity axiom that

$$U(\mathbf{r}^2) = U(\mathbf{r}) + U(\mathbf{r}) = 2U(\mathbf{r}) \qquad (A.9)$$

Secondly, by applying the Lemma A.1 we get

$$U(\mathbf{r}^2) = U(\mathbf{1}_{k^2}, \boldsymbol{\rho}_{n^2-k^2}) \qquad (A.10)$$
$$= U(\mathbf{1}_k, \boldsymbol{\rho}_{n^2-k})$$
$$+(1-\rho)U(\mathbf{1}_{k^2}, \mathbf{0}_{n^2-k^2})$$
$$-(1-\rho)U(\mathbf{1}_k, \mathbf{0}_{n^2-k})$$

We also know by the expansibility axiom that $U(\mathbf{r}) = U(\mathbf{r}, 0)$ for any possibility distribution $\mathbf{r}$ and we can apply Lemma A.1 to any possibility distribution and this will make the final q possibility values zero. Let us

L.3 deals with a special dist.

need add & expan

perform this operation upon the first term on the right hand side of the equality in Eq. (A.10),

$$
\begin{aligned}
U(\mathbf{1}_k, \boldsymbol{\rho}_{n^2-k}) &= U(\mathbf{1}_k, \boldsymbol{\rho}_{n-k}, \mathbf{0}_{n^2-n}) \\
&\quad + \rho U(\mathbf{1}_{n^2}) \\
&\quad - \rho U(\mathbf{1}_n, \mathbf{0}_{n^2-n})
\end{aligned}
$$

Substituting Eq. (A.7) into Eq. (A.10) we calculate that,

$$
\begin{aligned}
U(\mathbf{1}_{k^2}, \boldsymbol{\rho}_{n^2-k^2}) &= U(\mathbf{1}_k, \boldsymbol{\rho}_{n^2-k}) &\text{(A.11)} \\
&\quad + (1-\rho)U(\mathbf{1}_{k^2}, \mathbf{0}_{n^2-k^2}) \\
&\quad - (1-\rho)U(\mathbf{1}_k, \mathbf{0}_{n^2-k}) \\
&= U(\mathbf{1}_k, \boldsymbol{\rho}_{n-k}, \mathbf{0}_{n^2-n}) + \rho U(\mathbf{1}_{n^2}) - \rho U(\mathbf{1}_n, \mathbf{0}_{n^2-n})
\end{aligned}
$$

Now using the expansibility axiom to drop all final zeros and Lemma (A.2) to replace $U(\mathbf{1}_k)$ with $\log_2 k$,

$$
\begin{aligned}
U(\mathbf{1}_{k^2}, \boldsymbol{\rho}_{n^2-k^2}) &= U(\mathbf{1}_k, \boldsymbol{\rho}_{n-k}) + \rho \log_2 n^2 - \rho \log_2 n &\text{(A.12)} \\
&\quad + (1-\rho)\log_2 k^2 - (1-\rho)\log_2 n
\end{aligned}
$$

We now finish the proof by remembering that $U(\mathbf{1}_{k^2}, \boldsymbol{\rho}_{n^2-k^2}) = U(\mathbf{r}^2) = 2U(\mathbf{r})$, that $U(\mathbf{1}_k, \boldsymbol{\rho}_n) = U(\mathbf{r})$, and that $\log_2 x^2 = 2\log_2 x$:

$$
\begin{aligned}
2U(\mathbf{r}) &= U(\mathbf{r}) + 2\rho \log_2 n - \rho \log_2 n &\text{(A.13)} \\
&\quad + 2(1-\rho)\log_2 k - (1-\rho)\log_2 n \\
U(\mathbf{r}) &= \rho \log_2 n + (1-\rho)\log_2 k.
\end{aligned}
$$

∎

Theorem A.4 *The U-uncertainty is the only function that satisfies the axioms of expansibility, monotonicity, additivity, branching, and $U(1,1) = 1$.*

Proof. The proof is by induction on n, the length of the possibility distribution. If $n = 2$ then we can apply Lemma A.3 to get the desired result immediately. Assume now that the theorem is true for $n - 1$; then we can use the expansibility and branching axioms in the same way as was used in the proof of Lemma A.3 to replace r_n with zero, + math induction

$$
\begin{aligned}
U(\mathbf{r}) &= U(r_1, r_2, \ldots, r_n) &\text{(A.14)} \\
&= U(r_1, r_2, \ldots, r_n, 0) \\
&= U(r_1, r_2, \ldots, r_{n-1}, 0, 0) \\
&\quad + r_{n-1}U\left(\mathbf{1}_{n-1}, \frac{r_n}{r_{n-1}}, 0\right) \\
&\quad - r_{n-1}U(\mathbf{1}_{n-1}, 0, 0).
\end{aligned}
$$

We can now apply Lemma (A.3) to the term $U(\mathbf{1}_{n-1}, \frac{r_n}{r_{n-1}}, 0)$ and drop terminal zeros,

$$
\begin{aligned}
U(\mathbf{r}) &= U(r_1, r_2, \ldots, r_{n-1}) && \text{(A.15)}\\
&\quad + r_{n-1}\left[(1 - \frac{r_n}{r_{n-1}})\log_2(n-1) + \frac{r_n}{r_{n-1}}\log_2 n\right]\\
&\quad - r_{n-1}\log_2(n-1)\\
&= U(r_1, r_2, \ldots, r_{n-1})\\
&\quad + (r_{n-1} - r_n)\log_2(n-1) + r_n\log_2 n\\
&\quad - r_{n-1}\log_2(n-1)\\
&= U(r_1, r_2, \ldots, r_{n-1}) - r_n\log_2(n-1) + r_n\log_2 n\\
&= U(r_1, r_2, \ldots, r_{n-1}) + r_n\log_2\frac{n}{n-1}.
\end{aligned}
$$

Applying the induction hypothesis, we conclude

$$
\begin{aligned}
U(\mathbf{r}) &= U(r_1, r_2, \ldots, r_{n-1}) + r_n\log_2\frac{n}{n-1} && \text{(A.16)}\\
&= \sum_{i=2}^{n-1} r_i\log_2\frac{i}{i-1} + r_n\log_2\frac{n}{n-1}\\
&= \sum_{i=2}^{n} r_i\log_2\frac{i}{i-1}.
\end{aligned}
$$

BIBLIOGRAPHY

Aczél, J. and Z. Daróczy [1975], *On Measures of Information and Their Characterizations.* Academic Press, New York.

Aczél, J., B. Forte and C. T. Ng [1974], "Why the Shannon and Hartley entropies are 'natural'." *Advances in Applied Probability,* **6**, pp. 131-146.

Ash, R. B. [1965], *Information Theory.* Wiley-Interscience, New York.

Ashby, W. R. [1958], "Requiste variety and its implications for the control of complex systems." *Cybernetica,* **1**(2), pp. 83-99.

Ashby, W. R. [1965], "Measuring the internal informational exchange in a system." *Cybernetica,* **8**(1), pp. 5-22.

Ashby, W. R. [1969], "Two tables of identities governing information flows within large systems." *ASC Communications,* **1**(2), pp. 3-8.

Ashby, W. R. [1972], "Systems and their informational measures." In: Klir, G. J., ed., *Trends in General Systems Theory.* Wiley–Interscience, New York, pp. 78-97.

Avgers, T. G. [1983], "Axiomatic derivation of the mutual information principle as a method of inductive inference." *Kybernetes,* **12**(2), pp. 107-113.

Batten, D. F. [1983], *Spatial Analysis of Interacting Economics.* Kluwer-Nijhoff, Boston.

Bellman, R. E. and M. Giertz [1973], "On the analytic formalism of the theory of fuzzy sets." *Information Sciences,* 5, pp. 149-156.

Bharathi-Devi, B. and V. V. S. Sarma [1985], "Estimation of fuzzy memberships from histograms." *Information Sciences,* 35, pp. 43-59.

Billingsley, P. [1986], *Probability and Measure.* John Wiley, New York.

Blahut, R. E. [1987], *Principles and Practice of Information Theory.* Addison-Wesley, Reading, Massachusetts.

Bordley, R. F. [1983], "A central principle of science: optimization." *Behavioral Science,* 28(1), pp. 53-64.

Bremerman, H. J. [1962], "Optimization through evolution and recombination." In: Yovits, M. C., G. T. Jacobi and G. D. Goldstein, eds., *Self-Organizing Systems.* Spartan Books, Washington, pp. 93-106.

Brillouin, L. [1956], *Science and Information Theory.* Academic Press, New York.

Brillouin, L. [1964], *Scientific Uncertainty and Information.* Academic Press, New York.

Buck, B. and V. A. Macaulay [1991], *Maximum Entropy in Action.* Oxford University Press, Oxford and New York.

Cavallo, R. E. and G. J. Klir [1982], "Reconstruction of possibilistic behavior systems." *Fuzzy Sets and Systems,* 8(2), pp. 175-197.

Chaitin, G. J. [1987], *Information, Randomness, and Incompleteness: Papers on Algorithmic Information Theory.* World Scientific, Singapore.

Chellas, B. F. [1980], *Modal Logic: An Introduction.* Cambridge University Press, Cambridge and New York.

Cherry, C. [1957], *On Human Communication.* MIT Press, Cambridge, Massachusetts.

Christensen, R. [1980-81], *Entropy Minimax Sourcebook (4 volumes).* Entropy Limited, Lincoln, Massachusetts.

Christensen, R. [1985], "Entropy minimax multivariate statistical modeling–I: Theory." *Intern. J. of General Systems,* 11(3), pp. 231-277.

Christensen, R. [1986], "Entropy minimax multivariate statistical modeling–II: Applications." *Intern. J. of General Systems,* **12**(3), pp. 227-305.

Civanlar, M. R. and H. J. Trussell [1986], "Constructing membership functions using statistical data." *Fuzzy Sets and Systems,* **18**(1), pp. 1-13.

Cohen, L. J. [1970], *The Implications of Induction.* Methuen, London.

Conant, R. C. [1969], "The information transfer required in regulatory processes." *IEEE Trans. on Systems Sciences and Cybernetics,* **SSC-5**(4), pp. 334-338.

Conant, R. C. [1974], "Information flows in hierarchical systems." *Intern. J. of General Systems,* **1**(1), pp. 9-18.

Conant, R. C. [1976], "Laws of information which govern systems." *IEEE Trans. on Systems, Man, and Cybernetics,* **SMC-6**(4), pp. 240-255.

Conant, R. C. [1981], "Efficient proofs of identities in n-dimensional information theory." *Cybernetica,* **24**(3), pp. 191-197.

De Cooman, G. [1997], "Possibility Theory." *Intern. J. of General Systems,* **25**(1), I: pp. 291-323, II: pp. 325-351, III: pp. 353-371.

De Cooman, G., D. Ruan and E. E. Kerre, eds., [1995], *Foundations and Applications of Possibility Theory.* World Scientific, Singapore.

De Luca, A. and S. Termini [1972], "A definition of a nonprobabilistic entropy in the setting of fuzzy set theory." *Information and Control,* **20**(4), pp. 301-312.

De Luca, A. and S. Termini [1974], "Entropy of L-Fuzzy Sets." *Information and Control,* **24**(1), pp. 55-73.

De Luca, A. and S. Termini [1977], "On the convergence of entropy measures of a fuzzy set." *Kybernetes,* **6**(3), pp. 219-227.

Delgado, M. and S. Moral [1987], "On the concept of possibility-probability consistency." *Fuzzy Sets and Systems,* **21**(3), pp. 311-318.

Dempster, A. P. [1967a], "Upper and lower probabilities induced by a multivalued mapping." *Annals of Mathematical Statistics,* **38**(2), pp. 325-339.

Dempster, A. P. [1967b], "Upper and lower probability inferences based on a sample from a finite univariate population." *Biometrika,* **54** (3&4), pp. 515-528.

Denneberg, D. [1994], *Non-additive Measure and Integral.* Kluwer, Boston.

Devlin, K. [1991], *Logic and Information.* Cambridge University Press, Cambridge and New York.

Dombi, J. [1982], "A general class of fuzzy operators, the De Morgan class of fuzzy operators and fuzziness measures induced by fuzzy operators." *Fuzzy Sets and Systems,* 8(2), pp. 149-163.

Dretske, F. I. [1981], *Knowledge and the Flow of Information.* MIT Press, Cambridge, Massachusetts.

Dretske, F. I. [1983], "Presis of knowledge and the flow of information." *Behavioral and Brain Sciences,* 6(1), pp. 55-90.

Dubois, D. and H. Prade [1980a], *Fuzzy Sets and Systems: Theory and Applications.* Academic Press, New York.

Dubois, D. and H. Prade [1980b], "New results about properties and semantics of fuzzy set - theoretic operations." In: Wang, P. P. and S. K. Chang, eds., *Fuzzy Sets.* Plenum Press, New York, pp. 59-75.

Dubois, D. and H. Prade [1982], "On several representations of an uncertain body of evidence." In: Gupta, M. M. and E. Sanchez, eds., *Fuzzy Information and Decision Processes.* North-Holland, New York, pp. 167-181.

Dubois, D. and H. Prade [1985a], "A note on measures of specificity for fuzzy sets." *Intern. J. of General Systems,* 10(4), pp. 279-283.

Dubois, D. and H. Prade [1985b], "Evidence measures based on fuzzy information." *Automatica,* 21(5), pp. 547-562.

Dubois, D. and H. Prade [1985c], "Unfair coins and necessity measures: Towards a possibilistic interpretation of histograms." *Fuzzy Sets and Systems,* 10(1), pp. 15-20.

Dubois, D. and H. Prade [1986], "Fuzzy sets and statistical data." *European J. of Operational Research,* 25(3), pp. 345-356.

Dubois, D. and H. Prade [1987a], "An alternative approach to the handling of subnormal possibility distributions: A critical comment on a proposal by Yager." *Fuzzy Sets and Systems,* 24(1), pp. 123-126.

Dubois, D. and H. Prade [1987b], "Properties of measures of information in evidence and possibility theories." *Fuzzy Sets and Systems,* 24(2), pp. 161-182.

Dubois, D. and H. Prade [1987c], "The principle of minimum specificity as a basis for evidential reasoning." In: Bouchon, B. and R. R. Yager, eds., *Uncertainty in Knowledge-Based Systems*. Springer-Verlag, New York, pp. 75-84.

Dubois, D. and H. Prade [1988], *Possibility Theory: An Approach to Computerized Processing of Uncertainty*. Plenum Press, New York.

Dubois, D. and H. Prade [1990], "Consonant approximations of belief functions." *Intern. J. of Approximate Reasoning*, 4(5&6), pp. 419-449.

Dubois, D. and H. Prade [1991], "Fuzzy sets in approximate reasoning, Part I: Inference with possibility distributions." *Fuzzy Sets and Systems*, 40(1), pp. 143-202.

Dubois, D., H. Prade and S. Sandri [1991], "On possibility/probability transformations." In: *Proc. Fourth IFSA Congress: Mathematics*. Brussels, pp. 50-53

Ericson, G. J. and C. R. Smith [1988], *Maximum-Entropy and Bayesian Methods in Science and Engineering, Vol. 1: Foundations, Vol. II: Applications*. Kluwer, Dordrecht.

Feinstein, A. [1958], *Foundations of Information Theory*. McGraw-Hill, New York.

Frank, M. J. [1979], "On the simultaneous associativity of F(x, y) and x+y-F(x, y)." *Aequationes Mathematicae*, 19(2&3), pp. 194-226.

Garey, M. R. and D. S. Johnson [1979], *Computers and Intractability: A Guide to the Theory of NP-Completeness*. W. H. Freeman, San Francisco.

Geer, J. F. and G. J. Klir [1991], "Discord in possibility theory." *Intern. J. of General Systems*, 19(2), pp. 119-132.

Geer, J. F. and G. J. Klir [1992], "A mathematical analysis of information-preserving transformations between probabilistic and possibilistic formulations of uncertainty." *Intern. J. of General Systems*, 20(2), pp. 143-176.

Grabisch, M., H. T. Nguyen and E. A. Walker [1995], *Fundamentals of Uncertainty Calculi With Applications to Fuzzy Inference*. Kluwer, Boston.

Grandy, W. T. and L. H. Schick, eds., [1991], *Maximum Entropy and Bayesian Methods*. Kluwer, Boston.

Guan, J. W. and D. A. Bell [1991], *Evidence Theory and Its Applications, Vol. 1.* North-Holland, New York.

Guan, J. W. and D. A. Bell [1992], *Evidence Theory and Its Applications, Vol. 2.* North-Holland, New York.

Guiasu, S. [1977], *Information Theory with Applications.* McGraw-Hill, New York.

Gulati, S., J. Barhen and S. S. Iyengar [1991], "Neurocomputing formalisms for computational learning and machine intelligence." In: Yovits, M. C., ed., *Advances in Computers, Vol. 33.* Academic Press, San Diego, pp. 173-245.

Halmos, P. R. [1950], *Measure Theory.* D. Van Nostrand, Princeton, New Jersey.

Hamacher, H. [1978], "Über logische Verknüpfungen unscharfer Aussagen und deren zugehörige Bewertungs-funktionen." In: Trappl, R., G. J. Klir and L. Ricciardi, eds., *Progress in Cybernetics and Systems Research.* Hemisphere, Washington, D. C., pp. 276-287.

Hardy, G. H., J. E. Littlewood and G. Polya [1934], *Inequalities.* Cambridge University Press, Cambridge.

Harmanec, D. [1995], "Toward a characterization of uncertainty measure for the Dempster-Shafer theory." In: *Proc. of the Eleventh Intern. Conf. on Uncertainty in Artificial Intelligence.* Morgan-Kaufmann, San Mateo, California, pp. 255-261.

Harmanec, D. and G. J. Klir [1994], "Measuring total uncertainty in Dempster-Shafer theory: A novel approach." *Intern. J. of General Systems,* **22**(4), pp. 405-419.

Harmanec, D. and G. J. Klir [1996], "Principle of uncertainty revisited." In: *Proc. Fourth Intern. Fuzzy Systems and Intelligent Control Conf. (Maui, Hawaii, April 8-11, 1996),* pp. 331-339.

Harmanec, D. and G. J. Klir [1997], "On information-preserving transformations." *Intern. J. of General Systems,* **26**, pp. (to appear).

Harmanec, D., G. J. Klir and G. Resconi [1994], "On modal logic interpretation of Dempster-Shafer theory of evidence." *Intern. J. of Intelligent Systems,* **9**(10), pp. 941-951.

Harmanec, D., G. J. Klir and Z. Wang [1996], "Modal logic interpretation of Dempster-Shafer Theory: An infinite case." *Intern. J. of Approximate Reasoning,* **14**(2&3), pp. 81-93.

Harmanec, D., G. Resconi, G. J. Klir and Y. Pan [1996], "On the computation of uncertainty measure in Dempster-Shafer theory." *Intern. J. of General Systems,* **25**(2), pp. 153-163.

Hartley, R. V. L. [1928], "Transmission of information." *The Bell Systems Technical Journal,* **7**(3), pp. 535-563.

Henkind, S. J. and M. C. Harrison [1988], "An analysis of four uncertainty calculi." *IEEE Trans. on Systems, Man, and Cybernetics,* **18**(5), pp. 700-714.

Higashi, M. and G. J. Klir [1982], "On measures of fuzziness and fuzzy complements." *Intern. J. of General Systems,* **8**(3), pp. 169-180.

Higashi, M. and G. J. Klir [1983a], "Measures of uncertainty and information based on possibility distributions." *Intern. J. of General Systems,* **9**(1), pp. 43-58.

Higashi, M. and G. J. Klir [1983b], "On the notion of distance representing information closeness: Possibility and probability distributions." *Intern. J. of General Systems,* **9**(2), pp. 103-115.

Höhle, U. [1982], "Entropy with respect to plausibility measures." In: *Proc. 12th IEEE Intern. Symp. on Multiple-Valued Logic,* pp. 167-169.

Horvitz, E. J., D. E. Heckerman and C. P. Langlotz [1986], "A framework for comparing alternative formalisms for plausible reasoning." *Proc. Fifth National Conf. on Artificial Intelligence,* Philadelphia, pp. 210-214.

Hughes, G. E. and M. J. Cresswell [1968], *An Introduction to Modal Logic.* Methuen, London.

Hughes, G. E. and M. J. Cresswell [1984], *A Companion to Modal Logic.* Methuen, London and New York.

Jaynes, E. T. [1968], "Prior probabilities." *IEEE Trans. on Systems Science and Cybernetics,* **4**(3), pp. 227-241.

Jaynes, E. T. [1979], "Where do we stand on maximum entropy?" In: Levine, R. L. and M. Tribus, eds., *The Maximum Entropy Formalism.* MIT Press, Cambridge, Massachusetts, pp. 15-118.

Jaynes, E. T. [1982], "On the rationale of maximum entropy methods." *Proc. of IEEE,* **70**(9), pp. 939-952.

Jaynes, E. T. [1983], Rosenkrantz, R. D., ed., *Papers on Probability, Statistics and Statistical Physics.* D. Reidel, Dordrecht.

Jaynes, E. T. [1989], "Clearing up mysteries: The original goal." In: *Skilling, J., ed., Maximum Entropy and Bayesian Methods.* Kluwer, Dordrecht, pp. 1-27.

Joslyn, C. and G. J. Klir [1992], "Minimal information loss in possibilistic approximations of random sets." *Proc. IEEE Intern. Conf. on Fuzzy Systems,* San Diego, pp. 1081-1088.

Jumarie, G. [1994], "Possibility - probability transformation: A new result via information theory of deterministic functions." *Kybernetes,* **23**(5), pp. 56-59.

Jumarie, G. [1995], "Further results on possibility-probability conversion via relative information and informational invariance." *Cybernetics and Systems,* **26**(1), pp. 111-128.

Justice, J. H. [1986], *Maximum Entropy and Bayesian Methods in Applied Statistics.* Cambridge University Press, Cambridge.

Kandel, A. [1986], *Fuzzy Mathematical Techniques with Applications.* Addison Wesley, Reading, MA.

Kandel, A. and M. Schneider [1988], "Fuzzy sets and their applications to artificial intelligence." In: Yovits, M. C., ed., *Advances in Computers, Vol. 28.* Academic Press, San Diego, pp. 69-105.

Kapur, J. N. [1983], "Twenty-five years of maximum entropy principle." *J. Math. Phys. Sciences,* **17**(2), pp. 103-156.

Kapur, J. N. [1989], *Maximum Entropy Models in Science and Engineering.* John Wiley, New York.

Kapur, J. N. [1994], *Measures of Information and Their Applications.* John Wiley, New York.

Kapur, J. N. and H. K. Kesavan [1987], *The Generalized Maximum Entropy Principle (with Applications).* Sadford Educational Press, Waterloo, Canada.

Kaufmann, A. [1975], *Introdution to the Theory of Fuzzy Subsets - Vol. 1: Fundamental Theoretical Elements.* Academic Press, New York.

Kaufmann, A. and M. M. Gupta [1985], *Introduction to Fuzzy Arithmetic: Theory and Applications.* Van Nostrand, New York.

Klir, G. J. [1985], *Architecture of Systems Problem Solving.* Plenum Press, New York.

Klir, G. J. [1989a], "Is there more to uncertainty than some probability theorists might have us believe?" *Intern. J. of General Systems,* **15**(4), pp. 347-378.

Klir, G. J. [1989b], "Probability - possibility conversion." In: *Proc. Third IFSA Congress, Seattle,* pp. 408-411.

Klir, G. J. [1990a], "A principle of uncertainty and information invariance." *Intern. J. of General Systems,* **17**(2&3), pp. 249-275.

Klir, G. J. [1990b], "Dynamic aspects in reconstructability analysis: The role of minimum uncertainty principles." *Revue Internationale de Systemique,* **4**(1), pp. 33-43.

Klir, G. J. [1991a], *Facets of Systems Science.* Plenum Press, New York.

Klir, G. J. [1991b], "Generalized information theory." *Fuzzy Sets and Systems,* **40**(1), pp. 127-142.

Klir, G. J. [1991c], "Some applications of the principle of uncertainty invariance." *Proc. Intern. Fuzzy Eng. Symp.,* Yokohama, Japan, pp. 15-26.

Klir, G. J. [1993], "Developments in uncertainty-based information." In: Yovits, M. C., ed., *Advances in Computers, Vol. 36.* Academic Press, San Diego, pp. 255-332.

Klir, G. J. [1994a], "Multivalued logics versus modal logics: Alternative frameworks for uncertainty." In: Wang, P. P., ed., *Advances in Fuzzy Theory and Technology.* Duke Univ., Durham, N. C., pp. 3-47.

Klir, G. J. [1994b], "On the alleged superiority of probabilistic representation of uncertainty." *IEEE Trans. on Fuzzy Systems,* **2**(1), pp. 27-31.

Klir, G. J. [1995a], "From classical sets to fuzzy sets: A grand paradigm shift." In: Wang, P. P., ed., *Advances in Fuzzy Theory and Technology.* Duke University, Durham, pp. 5-30.

Klir, G. J. [1995b], "Principles of uncertainty: What are they? Why do we need them?" *Fuzzy Sets and Systems,* **74**(1), pp. 15-31.

Klir, G. J. and J. A. Cooper [1996], "On constraint fuzzy arithmetic." In: *Proc. Fifth IEEE Intern. Conf. on Fuzzy Systems,* pp. 1693-1699.

Klir, G. J. and T. Folger [1988], *Fuzzy Sets, Uncertainty, and Information.* Prentice Hall, Englewood Cliffs, NJ.

Klir, G. J. and D. Harmanec [1994], "On modal logic interpretation of possibility theory." *Intern. J. of Uncertainty, Fuzziness, and Knowledge-Based Systems,* **2**(2), pp. 237-245.

Klir, G. J. and D. Harmanec [1995], "On some bridges to possibility theory." In: De Cooman, G. and et. al., eds., *Foundations and Applications of Possibility Theory.* World Scientific, Singapore, pp. 3-19.

Klir, G. J. and M. Mariano [1987], "On the uniqueness of possibilistic measure of uncertainty and information." *Fuzzy Sets and Systems,* **24**(2), pp. 197-219.

Klir, G. J., M. Mariano, M. Pittarelli, and K. Kornwachs [1988], "The potential of reconstructability analysis for production research." *Intern J. of Production Research,* **26**(4), pp. 629-645.

Klir, G. J. and B. Parviz [1986], "General reconstruction characteristics of probabilistic and possibilistic systems." *Intern. J. of Man-Machine Systems,* **25**(4), pp. 367-397.

Klir, G. J. and B. Parviz [1992a], "A note on the measure of discord." In: Dubois, D., ed., *Proceedings of the Eighth Conference on Artificial Intelligence.* Morgan Kaufmann, San Mateo, California, pp. 138-141.

Klir, G. J. and B. Parviz [1992b], "Probability-possibility transformations: A comparison." *Intern. J. of General Systems,* **21**(3), pp. 291-310.

Klir, G. J., B. Parviz and M. Higashi [1986], "Relationship between true and estimated possibilistic systems and their reconstruction." *Intern. J. of General Systems,* **12**(4), pp. 319-331.

Klir, G. J. and A. Ramer [1990], "Uncertainty in the Dempster-Shafer theory: A critical re-examination." *Intern. J. of General Systems,* 18(2), pp. 155-166.

Klir, G. J. and E. C. Way [1985], "Reconstructablility analysis: Aims, results, open problems." *Systems Research,* **2**(2), pp. 141-163.

Klir, G. J. and M. Wierman [1987], "On properties of the V–uncertainty." In: Chameau, J. L. and J. T. P. Yau, eds., *Proc. 1987 NAFIPS Workshop,* pp. 96-106.

Klir, G. J. and B. Yuan [1993], "On measures of conflict among set-valued statements." *Proc. 1993 World Congress on Neural Networks; July 11-15, 1993, Portland, Oregon,* pp. 627-630.

Klir, G. J. and B. Yuan [1995a], *Fuzzy Sets and Fuzzy Logic: Theory and Applications.* Prentice Hall, Upper Saddle River, NJ.

Klir, G. J. and B. Yuan [1995b], "On nonspecificity of fuzzy sets with continuous membership functions." In: *Proc. 1995 Intern. Conf. on Systems, Man, and Cybernetics,* Vancouver, pp. 25-29.

Klir, G. J. and B. Yuan [1996], *Fuzzy Sets, Fuzzy Logic, and Fuzzy Systems: Selected Papers by Lotfi A. Zadeh.* World Scientific, Singapore.

Kohlas, J. and P. A. Monney [1995], *A Mathematical Theory of Hints: An Approach to the Dempster-Shafer Theory of Evidence.* Springer, Berlin.

Kolmogorov, A. N. [1965], "Three approaches to the quantitative definition of information." *Problems of Information Transmission,* **1**(1), pp. 1-7.

Kosko, B. [1993], "Addition as fuzzy mutual entropy." *Information Sciences,* **73**(3), pp. 273-284.

Kuhn, T. S. [1962], *The Structure of Scientific Revolutions.* Univ. of Chicago Press, Chicago.

Kullback, S. [1959], *Information Theory and Statistics.* John Wiley, New York.

Kyburg, H. E. [1987], "Bayesian and non-Bayesian evidential updating." *Artificial Intelligence,* **31**(3), pp. 271-293.

Lamata, M. T. and S. Moral [1988], "Measures of entropy in the theory of evidence." *Intern. J. of General Systems,* **14**(4), pp. 297-305.

Lee, N. S., Y. L. Grize and K. Dehnad [1987], "Quantitative models for reasoning under uncertainty in knowledge-based expert systems." *Intern. J. of Intelligent Systems,* **2**(1), pp. 15-38.

Leung, Y. [1982], "Maximum entropy estimation with inexact information." In: Yager, R. R., ed., *Fuzzy Set and Possibility Theory.* Pergamon Press, Oxford, pp. 32-37.

Levine, R. D. and M. Tribus, eds., [1979], *The Maximum Entropy Formalism.* MIT Press, Cambridge, Massachusetts.

Lowen, R. [1996], *Fuzzy Set Theory: Basic Concepts, Techniques and Bibliography.* Kluwer, Boston.

Maeda, Y. and H. Ichihashi [1993], "An uncertainty measure with monotonicity under the random set inclusion." *Intern. J. of General Systems,* **21**(4), pp. 379-392.

Maeda, Y., H. T. Nguyen and H. Ichihashi [1993], "Maximum entropy algorithms for uncertainty measures." *Intern. J. of Uncertainty, Fuzziness and Knowledge-Based Systems,* **1**(1), pp. 69-93.

Mariano, M. J. [1987], *Aspects of Inconsistency in Reconstructability Analysis.* Ph. D. Dissertation, SUNY-Binghamton, New York.

Mathai, A. M. and Rathie P. N. [1975], *Basic Concepts of Information Theory and Statistics.* John Wiley, New York.

Matheron, G. [1975], *Random Sets and Integral Geometry.* John Wiley, New York.

Maung, I. [1995], "Two characterizations of a minimum-information principle for possibilistic reasoning." *Intern. J. of Approximate Reasoning,* **12**(2), pp. 133-156.

McLaughlin, D. W., ed., [1984], *Inverse Problems.* American Mathematical Society, Providence, Rhode Island.

Meyerowitz, A., F. Richman and E. A. Walker [1994], "Calculating maximum-entropy probability densities for belief functions." *Intern. J. of Uncertainty, Fuzziness, and Knowledge-Based Systems,* **2**(4), pp. 377-389.

Moore, R. E. [1966], *Interval Analysis.* Prentice Hall, Englewood Cliffs, New Jersey.

Moore, R. E. [1979], *Methods and Applications of Interval Analysis.* SIAM, Philadelphia.

Moral, S. [1986], "Construction of a probability distribution from a fuzzy information." In: Jones, A., A. Kaufmann and H. J. Zimmermann, eds., *Fuzzy Sets Theory and Applications.* D. Reidel, Dordrecht, pp. 51-60.

Nguyen, H. T. and E. A. Walker [1997], *A First Course in Fuzzy Logic.* CRC Press, Boca Raton, Florida.

Novák, V. [1989], *Fuzzy Sets and Their Applications.* Adam Hilger, Bristol, U. K.

Padet, C. [1996], "On applying information principles to fuzzy control." *Kybernetes,* **25**(1), pp. 61-64.

Pap, E. [1995], *Null-Additive Set Functions.* Kluwer, Boston.

Paris, J. B. and A. Vencovska [1989], "On the applicability of maximum entropy to inexact reasoning." *Intern. J. of Approximate Reasoning*, **3**(1), pp. 1-34.

Paris, J. B. and A. Vencovska [1990], "A note on the inevitability of maximum entropy." *Intern. J. of Approximate Reasoning*, **4**(3), pp. 183-224.

Pawlak, Z. [1991], *Rough Sets: Theoretical Aspects of Reasoning About Data.* Kluwer, Boston.

Ramer, A. [1987], "Uniqueness of information measure in the theory of evidence." *Fuzzy Sets and Systems*, **24**(2), pp. 183-196.

Ramer, A. [1990a], "Axioms of uncertainty measures: Dependence and independence." *Fuzzy Sets and Systems*, **35**(2), pp. 185-196.

Ramer, A. [1990b], "Information measures for continuous possibility distributions." *International Journal of General Systems*, **24**(2), pp. 183-196.

Ramer, A. and G. J. Klir [1993], "Measures of discord in the Dempster-Shafer theory." *Information Sciences*, **67**(1&2), pp. 35-50.

Ramer, A. and L. Lander [1987], "Classification of possibilistic uncertainty and information functions." *Fuzzy Sets and Systems*, **24**(2) , pp. 221-230.

Rényi, A. [1970], Probability Theory. North-Holland, Amsterdam.

Resconi, G., G. J. Klir, D. Harmanec and U. St. Clair [1996], "Interpretations of various uncertainty theories using models of modal logic: A summary." *Fuzzy Sets and Systems*, **80**(1), pp. 7-14.

Resconi, G., G. J. Klir, U. St. Clair and D. Harmanec [1993], "On the integration of uncertainty theories." *Intern. J. of Uncertainty, Fuzziness, and Knowledge-Based Systems*, **1**(1), pp. 1-18.

Resconi, G., G. J. Klir and U. St. Clair [1992], "Hierarchical uncertainty metatheory based upon modal logic." *Intern. J. of General Systems*, **21**(1), pp. 23-50.

Reza, F. M. [1961], *Introduction to Information Theory.* McGraw-Hill, New York.

Schweizer, B. and A. Sklar [1983], *Probability Metric Spaces.* North-Holland, New York.

Shackle, G. L. S. [1961], *Decision, Order and Time in Human Affairs.* Cambridge University Press, Cambridge.

Shackle, G. L. S. [1979], *Imagination and the Nature of Choice.* Edinburgh University Press, Edinburgh.

Shafer, G. [1976], *A Mathematical Theory of Evidence.* Princeton University Press, Princeton, New Jersey.

Shafer, G. [1985], "Belief functions and possibility measures." In: Bezdek, J. C., ed., *Analysis of Fuzzy Information.* CRC Press, Boca Raton, Florida, pp. 51-84.

Shafer, G. [1990], "Perspectives on the theory and practice of belief functions." *Intern. J. of Approximate Reasoning,* **4**(5&6), pp. 323-362.

Shannon, C. E. [1948], "The mathematical theory of communication." *The Bell System Technical Journal,* **27**(3&4), pp. 379-423, 623-656.

Shannon, C. E. and W. Weaver [1949], *The Mathematical Theory of Communication.* University of Illinois Press, Urbana, Ill.

Shore, J. E. and R. W. Johnson [1980], "Axiomatic derivation of the principle of maximum entropy and the principle of minimum cross-entropy." *IEEE Trans. on Information Theory,* **IT-26**(1), pp. 26-37.

Shore, J. E. and R. W. Johnson [1981], "Properties of cross-entropy minimization." *IEEE Trans. on Information Theory,* **27**(4), pp. 472-482.

Skilling, J. [1989], *Maximum-Entropy and Bayesian Methods.* Kluwer, Dordrecht.

Smets, P. [1988], "Belief functions." In: Smets, P., et al., eds., *Non-Standard Logics for Automated Reasoning.* Academic Press, San Diego, pp. 253-286.

Smets, P. [1990a], "Constructing the pignistic probability function in a context of uncertainty." In: Henrion et al., eds., *Uncertainty in AI 5.* North-Holland, New York, pp. 29-39.

Smets, P. [1990b], "The combination of evidence in the transferable belief model." *IEEE Trans. on Pattern Analysis and Machine Intelligence,* **12**(5), pp. 447-458.

Smith, C. R. and G. J. Ericson, eds., [1987], *Maximum-Entropy and Bayesian Spectral Analysis and Estimation Problems.* Reidel, Boston.

Smith, C. R., G. J. Ericson and P. O. Neudorfer, eds., [1992], *Maximum Entropy and Bayesian Methods.* Kluwer, Boston.

Smith, C. R. and W. T. Grandy, eds., [1985], *Maximum-Entropy and Bayesian Methods in Inverse Problems.* D. Reidel, Boston.

Smith, S. A. [1974], "A derivation of entropy and the maximum entropy criterion in the context of decision problems." *IEEE Trans. on Systems, Man, and Cybernetics,* **SMC-4**(2), pp. 157-163.

Smithson, M. [1989], *Ignorance and Uncertainty: Emerging Paradigms.* Springer-Verlag, New York.

Stephanou, H. E. and A. P. Sage [1987], "Perspectives on imperfect information." *IEEE Transactions on Systems, Man, and Cybernetics,* **17**(5), pp. pp. 780-798.

Stonier, T. [1990], *Information and the Internal Structure of the Universe.* Springer-Verlag, New York.

Strat, T. M. [1990], "Decision analysis using belief functions." *Intern. J. of Approximate Reasoning,* 4(5&6), pp. 391-417.

Strat, T. M. and J. D. Lowrance [1989], "Explaining evidential analysis." *Intern. J. of Approximate Reasoning,* **3**(4), pp. 299-353.

Sugeno, M. [1974], *Theory of Fuzzy Integrals and its Applications.* Ph. D. Dissertation, Tokyo Institute of Technology, Tokyo.

Sugeno, M. [1977], "Fuzzy measures and fuzzy integrals: A survey." In: Gupta, M. M., G. N. Saridis and B. R. Gaines, eds., *Fuzzy Automata and Decision Processes.* North-Holland, Amsterdam and New York, pp. 89-102.

Tarantola, A. [1987], *Inverse Problem Theory.* Elsevier, New York.

Theil, H. [1987], *Economics and Information Theory.* North-Holland, Amsterdam, and Rand McNally, Chicago.

Theil, H. and D. G. Fiebig [1984], *Exploiting Continuity: Maximum Entropy Estimation of Continuous Distributions.* Ballinger, Cambridge, Massachusetts.

Tribus, M. [1969], *Rational Descriptions, Decisions and Designs.* Pergamon Press, Oxford.

Vejnarová, J. and G. J. Klir [1993], "Measure of strife in Dempster-Shafer theory." *Intern. J. of General Systems,* **22**(1), pp. 25-42.

Voorbraak, F. [1989], "A computationally efficient approximation of Dempster-Shafer theory." *Intern. J. of Man-Machine Studies,* **30**(5), pp. 525-536.

Walley, P. [1991], *Statistical Reasoning With Imprecise Probabilities.* Chapman and Hall, London.

Walley, P. and T. L. Fine [1979], "Varieties of model (classificatory) and comparative probability." *Synthese,* **41**(3), pp. 321-374.

Wang, Z. and G. J. Klir [1992], *Fuzzy Measure Theory.* Plenum Press, New York.

Wang, Z., G. J. Klir and G. Resconi [1995], "Expressing fuzzy measure by a model of modal logic: A discrete case." In: Bien, Z. and K. C. Min, eds., *Fuzzy Logic and Its Applications to Engineering, Information Sciences, and Intelligent Systems.* Kluwer, Boston, pp. 3-13.

Watanabe, S. [1981], "Pattern recognition as a quest for minimum entropy." *Pattern recognition,* **13**(2), pp. 381-387.

Watanabe, S. [1985], *Pattern Recognition: Human and Mechanical.* John Wiley, New York.

Weaver, W. [1948], "Science and complexity." *American Scientist,* **36**(4), pp. 536-544.

Webber, M. J. [1979], *Information Theory and Urban Spatial Structure.* Croom Helm, London.

Whittemore, B. J. and M. C. Yovits [1973], "A generalized conceptual development for the analysis and flow of information." *J. of the American Society for Information Sciences,* **24**(3), pp. 221-231.

Whittemore, B. J. and M. C. Yovits [1974], "The quantification and analysis of information used in decision processes." *Information Sciences,* **7**(2), pp. 171-184.

Wierman, M. J. [1993], *Possibilistic Image Processing.* Ph. D. Dissertation, SUNY-Binghamton, New York.

Williams, P. M. [1980], "Bayesian conditionalisation and the principle of minimum information." *British J. for the Philosophy of Science,* **31**(2), pp. 131-144.

Wilson, A. G. [1970], *Entropy in Urban and Regional Modelling.* Pion, London.

Wonneberger, S. [1994], "Generalization of an invertible mapping between probability and possibility." *Fuzzy Sets and Systems,* **64**(2), pp. 229-240.

Yager, R. R. [1979], "On the measure of fuzziness and negation. Part I: Membership in the unit interval." *Intern. J. of General Systems,* **5**(4), pp. 189-200.

Yager, R. R. [1980a], "On a general class of fuzzy connectives." *Fuzzy Sets and Systems,* **4**(3), pp. 235-242.

Yager, R. R. [1980b], "On the measure of fuzziness and negation. Part II: Lattices." *Information and Control,* **44**(3), pp. 236-260.

Yager, R. R. [1983], "Entropy and specificity in a mathematical theory of evidence." *Intern. J. of General Systems,* **9**(4), pp. 249-260.

Yager, R. R. [1984], "On different classes of linguistic variables defined via fuzzy subsets." *Kybernetes,* **13**(2), pp. 103-110.

Yager, R. R. [1986], "A modification of the certainty measure to handle subnormal distributions." *Fuzzy Sets and Systems,* **20**(3), pp. 317-324.

Yager, R. R. [1987], "Toward a theory of conjunctive variables." *Intern. J. of General Systems,* **13**(3), pp. 203-227.

Yager, R. R. [1988], "Reasoning with conjunctive knowledge." *Fuzzy Sets and Systems,* **28**(1), pp. 69-83.

Yager, R. R., S. Ovchinnikov, R. M. Tong and H. T. Nguyen, eds., [1987], *Fuzzy Sets and Applications-Selected Papers by L. A. Zadeh.* John Wiley, New York.

Yen, J. [1989], "Gertis: a Dempster-Shafer approach to diagnosing hierarchical hypotheses." *ACM Communications,* **32**(5), pp. 573-585.

Yen, J. [1990], "Generalizing the Dempster-Shafer theory to fuzzy sets." *IEEE Transactions on Systems, Man, and Cybernetics,* **20**(3), pp. 559-570.

Yoo, M. [1994], *Information-Sensitive Fuzzy Database System for Decision Making and Control Using Information Invariance Principle.* Ph. D. Dissertation, SUNY-Binghamton, New York.

Yovits, M. C., C. R. Foulk and L. L. Rose [1981], "Information flow and analysis: Theory, simulation, and experiments." *J. of the American Society for Information Science,* **32**(3&4), pp. 187-210, 243-248.

Zadeh, L. A. [1965], "Fuzzy sets." *Information and Control,* **8**(3), pp. 338-353.

Zadeh, L. A. [1978], "Fuzzy sets as a basis for a theory of possibility." *Fuzzy Sets and Systems,* **1**(1), pp. 3-28.

Zadeh, L. A. [1979], "Fuzzy Sets and Information Granularity." In: Gupta, M. M., R. K. Ragade and R. R. Yager, eds., *Advances in Fuzzy Set Theory and Applications.* North-Holland, New York, pp. 3-18.

Zimmermann, H. J. [1985], *Fuzzy Set Theory-and Its Applications.* Kluwer, Boston.

INDEX

Studies in Fuzziness and Soft Computing